The Corporate Environmental Leader:

Five Steps to a New Ethic

By Twyla Dell

THE CORPORATE ENVIRONMENTAL LEADER

Twyla Dell

CREDITS
Managing Editor: **Kathleen Barcos**
Editor: **Kay Kepler**
Designer: **ExecuStaff**
Typesetting: **ExecuStaff**
Cover Design: **London Road Design**

Printed in the United States of America by Bawden Printing Company.

English language Crisp books are distributed worldwide. Our major international distributors include:

CANADA: Reid Publishing Ltd., Box 69559–109 Thomas St., Oakville, Ontario, Canada L6J 7R4. TEL: (905) 842-4428, FAX: (905) 842-9327

Raincoast Books Distribution Ltd., 112 East 3rd Avenue, Vancouver, British Columbia, Canada V5T 1C8. TEL: (604) 873-6581, FAX: (604) 874-2711

AUSTRALIA: Career Builders, P.O. Box 1051, Springwood, Brisbane, Queensland, Australia 4127. TEL: 841-1061, FAX: 841-1580

NEW ZEALAND: Career Builders, P.O. Box 571, Manurewa, Auckland, New Zealand. TEL: 266-5276, FAX: 266-4152

JAPAN: Phoenix Associates Co., Mizuho Bldg. 2-12-2, Kami Osaki, Shinagawa-Ku, Tokyo 141, Japan. TEL: 3-443-7231, FAX: 3-443-7640

Selected Crisp titles are also available in other languages. Contact International Rights Manager Suzanne Kelly at (415) 323-6100 for more information.

Library of Congress Catalog Card Number 93-72978
Dell, Twyla
The Corporate Environmental Leader
ISBN 1-56052-253-4

About the Book

The *Corporate Environmental Leader* tells us about yesterday's environmental disasters and today's environmental successes. Profiles of thirty of America's corporate leaders disclose how much time and resources they spend on environmental solutions, new technologies, community outreach and ecological restoration. Personal stories of small changes making big differences add further inspiration. New principles for doing business add guidance for today's corporate environmental leaders who already make a difference.

Acknowledgments

I want to thank the many friendly, hardworking, and environmentally enthusiastic people within the corporations I surveyed. Their courtesy and eagerness to share their programs and assist with information was heartening and much appreciated. Even more, their dedication to such programs as zero waste, new technologies, partnerships with unexpected allies to create solutions and to restore some of our natural fabric was inspiring. We should all be encouraged and look forward to even greater achievements.

I would also like to thank the telephone reference librarians at the Johnson County, Kansas, library system. They fielded every kind of question over the course of three months from ancient history to today's environmental laws and never failed me once.

Dedication

To my mother who loved the natural world,
and to my father who loved the world of business.

Contents

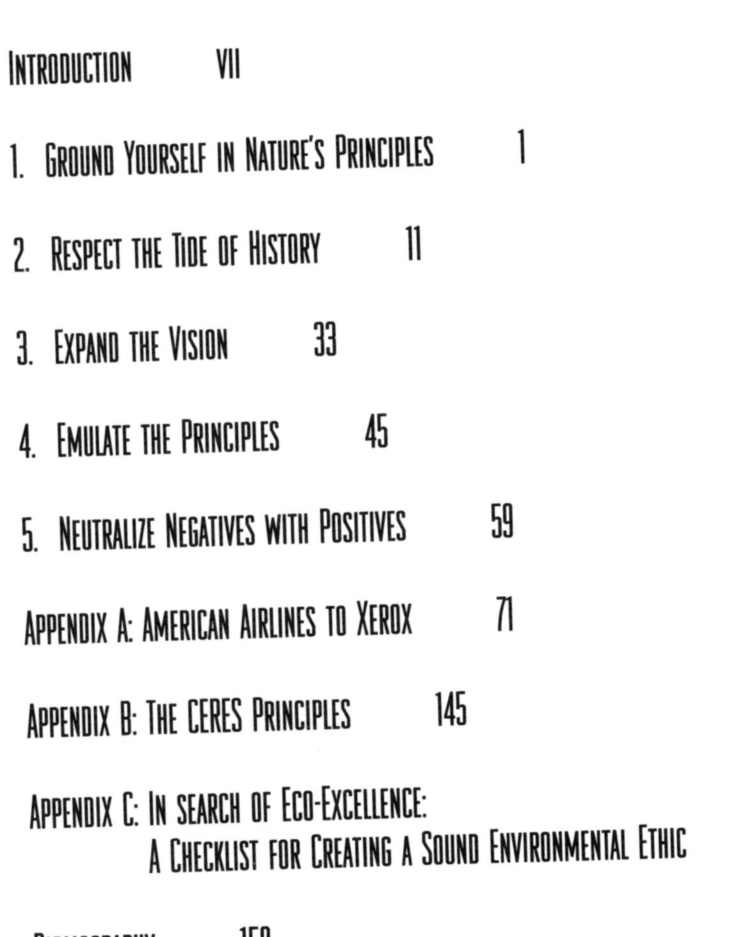

INTRODUCTION

The corporate environmental leader knows we are connected to a larger ecology than the organization alone.

Even five years ago a phrase such as "corporate environmental leader" would have sounded like an oxymoron. Now it is not only less a contradiction in terms, but an emerging role and market position. In fact, someone said to me not long ago, "A hundred years from now, businesses will be so green you'll be able to squeeze one and chlorophyll will run out." While we had a good laugh over such an extravagant statement, I wondered just what "green" might look like, and how close we were to getting there. Who knows? Business might turn just that—green.

The role of corporations in restoring and maintaining the environment is changing dramatically. Often painted as environmental villains, such organizations often now see their public responsibilities more clearly. This shift in roles and responsibilities leads us to the question of this book: Can corporations become environmental leaders?

If the answer is yes, what must happen for that corporate body to become an environmental leader? Can organizations save money by becoming eco-efficient? Can they afford *not* to do their best by the earth? Do they have any environmental responsibility to their communities beyond compliance with regulations? Can environmental technology become a market position? Can workers change their employers' behavior? If so, how do they do it?

These questions of the 1990s mark a new connection between business and the environment. This connection is the result of a simple understanding basic to our quality of life since early times: We depend on nature's bounty for our own. In 20th-century terms, an exhausted earth means an exhausted world economy. We have managed earth's harvest for many thousands of years, and we've had great successes and great failures. As we approach the commercialization of the entire globe, we also have begun to realize that the earth will be managed by humans just so far, then it will manage us.

A growing awareness of environmental history and modern ecological disasters has helped us to reevaluate our approach to harvesting earth's bounty—a task increasingly assumed by the corporate world. What we do in the next few years to protect, maintain and enhance our resources will make all the difference in the quality of life for ourselves and for future generations.

The purpose of this book is to give emerging corporate environmental leaders a sense of connection with nature's systems and a connection with their environmental past so mistakes won't be copied. The second purpose is to provide a photo album, so to speak, of where the American corporate world is in the mid-1990s in terms of environmental leadership.

I invite all corporate environmental leaders to demonstrate how the economy and ecology reinforce each other in a natural balance. Each small effort moves us forward. Join in; there is no turning back.

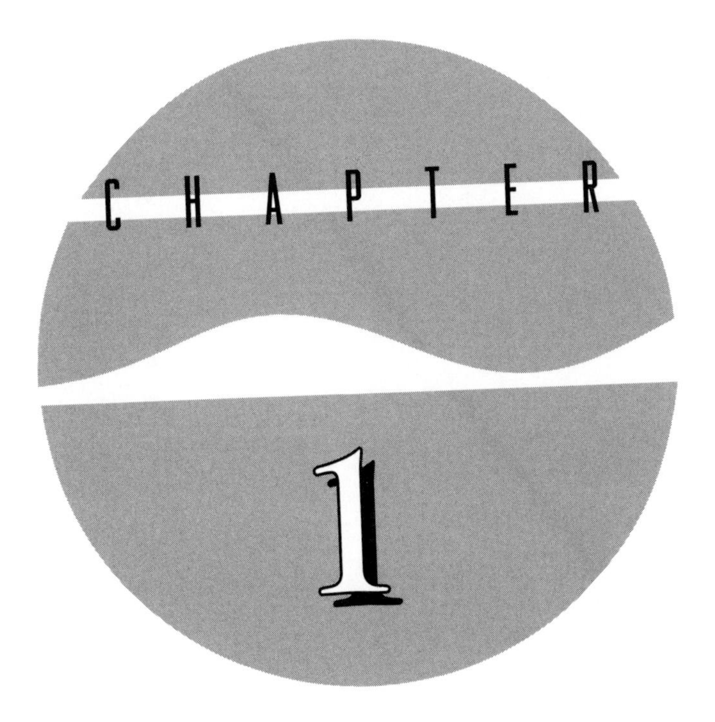

The corporate environmental leader makes friends with the planet.

Ground Yourself In Nature's Principles

Nature has seasons and cycles so rhythmic we lull ourselves into ignoring them. Although these rhythms weave the very threads of life on this planet, we have immunized ourselves against them—deeply, willfully and thoroughly. In the artificial world we have created for ourselves, only the great and dramatic acts of nature intrude—earthquakes, hurricanes, forest fires, floods, volcanic eruptions. These get our attention. But of the daily acts of nature, the sextillion daily threads woven together to provide us life, we are oblivious.

With little or no thought toward nature we barely notice forces far greater than our most powerful hydrogen bombs at work daily on this planet. We participate in these forces without knowing that we're playing in a league way over our heads. Look at these connections.

- Oceans speak to skies. Rain falls in Wichita—or not, depending on this conversation.
- Power-plant stacks spew their dew in Ohio. Depending on the wind, sugar maple trees in Vermont wither.
- City air conditioners leak atoms from millions of humming units. Frogs in mountain streams lose their skins to sunburn from a thinning ozone layer.

- Teakwood adorns fine offices as rainforests die to the gnashing teeth of big-business chainsaws. The moisture once held by green towering giants dries up. Streambeds slow to a trickle. Clouds move away. Remaining wildlife lacks cover to grow, and so it starves and dies out. Aborigines, yanked from their leafy lives, stream to the concrete landscapes of cities to a life of poverty and despair.

- Millions of automobiles cruise the planet. Billions of tons of pollution belch into the daily air. The blue sky turns brown. The atmosphere warms. The ocean notices, says something new to the sky which quits raining on Wichita while the waves rise on Miami.

Nature is so diverse that biologists estimate rainforest destruction makes at least one species of plant or animal life become extinct every day before it can be named and catalogued, with untold effect on the world's ecosystem. The question we now ask ourselves is this: How shall we live and do business in a way that honors rather than ignores that complexity? Is it a goal even worth striving for? If we were to live sustainably, how would we live? Where do we begin?

If we observe the signals nature gives us, we can learn the principles upon which life on this planet operates, and then we might reinvent ourselves using nature as the model. Try on some of nature's simple truths.

Nature is a renewable system

The first page of the book of nature trumpets a simple but astounding truth: In nature everything is biodegradable. Nature offers a far more sophisticated system than the one-way, use-it-and-toss-it life we have designed. It recycles everything.

Nature wears a seamless garment. It may look like a tiger here, a fish there, a tree yonder, an ocean beyond, the sky above, but each is a part of an intricate working system in which the waste of one creature is the food of another. Nothing is left unused. American conservationist Aldo Leopold calls that the "biotic pyramid."

Plants absorb energy from the sun. This energy flows through a circuit called the biota which maybe represented by a pyramid consisting of layers. The bottom layer is soil. . . . The species of a layer are alike not in where they came from or in what they look like, but in what they eat. Each successive layer depends on those below for food and often for other services, and each in turn furnishes food and services to those above. (A Sand County Almanac)

Creatures, strung together by food chains, produce no mountain of garbage, no odious landfills. Leopold says that the "tangle of chains [is] so complex as to seem disorderly, yet the stability of the system proves it to be a highly organized structure," and at the top of the chains emerge such miracles of evolution as the grizzly bear, blue whale, bald eagle, Bengal tiger, redwood forest, sea of prairie grass, coral reef—each living jewel "the spire of an edifice abuilding since the morning stars sang together."

Nature is a community, not a commodity

We have turned the natural world into the material world and called it progress. We must now ask: progress toward what? We have fouled our air, water and land. If we cannot sustain the natural world, we cannot sustain ourselves, because an economic system works by recirculating its wealth. We need to recirculate nature's wealth, because we are citizens of the biological community, not consumers of it.

Nature creates no hazardous waste

Nuclear waste is toxic for 200,000 years. Without special equipment, people who handle it die. Storing it safely is an enormous problem, because leaks in storage containers could contaminate the earth, air, or oceans—wherever it is kept—and kill thousands of people and injure thousands more.

A potential site that would bury as much as 70,000 tons of radioactive waste in Yucca Mountain, Nevada, could be ready by 2010. The nuclear power industry, however, will have produced more than 70,000 tons of waste before 2020, a few years after that site might open. Moreover, just getting the waste to the site would entail risk. The waste is now stored at the aging nuclear power plants where it is created. No solution rises over the horizon.

Air pollution can cause asthma, emphysema, and other breathing problems and related illnesses, which can kill or reduce life expectancy rates for those it smothers. In the United States, 22 million tons of sulfur dioxide emissions and 60 million tons of carbon monoxide emissions are poured into the air every year. Add to this cocktail 50–200 million pounds each of the top 10 toxic chemicals released each year and a million tons of the carbofluorocarbons (CFC) chemicals manufactured every year and put to work in 100 million refrigerators, 30 million freezers, 45 million home air conditioners, 50 million car air conditioners and millions more commercial units in the United States alone. This picture shows a manipulation of natural systems that make them unnatural, nonbiodegradable, toxic and finally unusable to the natural energy system that powers the earth.

Nature has no boundaries, but it does have limits

The earth is populated by 5.5 billion people, a number that is expected to double in the next century. The demands of civilization, especially Western industrial capitalism, strain the earth's fragile systems beyond return. The citizens of the West enjoy the highest living standards of the world, which entails enormous energy production and consumption in the construction and maintenance of automobiles, appliances and other consumer goods.

Industrial expansion can be blamed for the extinction of plant and animal species. The American passenger pigeon, for instance, was once considered to be the most numerous bird on the planet. Estimates ranged to five billion birds when America was first colonized. That's as many people as we now have on the planet. Imagine the flocks! When the birds flew overhead, they darkened the sky for hours. When they landed on trees, the weight of their collective bodies broke the branches. All a hunter had to do was point a shotgun skyward and a generous supply of food fell like manna from the sky.

In the space of 300 years these birds were obliterated by hunting. At first, colonists used them as a food supply. When the railroads were built in the 1800s, the Western states could ship barrel loads of birds to the East, which had been hunted out. Hundreds of thousands of birds were killed and shipped each year until they became too scarce to hunt. The last passenger pigeon died in the Cincinnati Zoo in 1914. Nature has no boundaries, but it does have limits.

Nature flows with continuous energy

All living things are part of the life energy system of the planet. As the system's members die off, the interconnections break down. Humans cannot live alone on a planet without the infinite strands of life energy supporting it. We must protect this system of which each species, including ourselves, is a part. Such vigilance is no more or less than protecting our capital, our natural capital. Any seasoned businessperson knows the value of capital and the temptation to spend it and then borrow back—exactly what we've been doing both ecologically and economically.

We can't possibly know nature's complexity and choose which species, which ecosystem, which biota to keep and which to destroy as not important. We must protect them all. Despite the gargantuan size of the planet, the human onslaught against it is greater than the biosphere and atmosphere can absorb.

We thought we were creating energy when we burned fossil fuel, but the flow of energy around the planet has not as much to do with fossil-fuel energy as it does with biotic energy, the ecosystem's life cycles. Leopold reminds us: "The current is the stream of energy which flows out of the soil into plants, thence into animals, thence back into the soil in a never-ending circuit of life."

That "circuit of life" constitutes our most precious possession. That's the money in nature's bank. Nature's output is our input. Our output, then, must become nature's input for that current of energy to complete its cycle. We cannot survive on a "civilized" planet covered in concrete, sustained by farm factories of chickens, cattle and hogs, the remaining wild creatures locked in small zoos. The planet can't survive without the myriad organisms that pass energy along the food chains up and down the endless strands of the web of life that create our ecosphere.

The ecosphere forms that foundation of all we do on this planet. On that foundation we have built an econosphere—a network of economic activity that derives all its energy and raw materials from the ecosphere. The economy, then, can be only as strong as the ecosphere is healthy and plentiful. Any good businessperson can see that the econosphere depends on the ecosphere for its existence.

Right now the rules for the business world tell us to make a profit first and clean up our messes second. In the new paradigm we won't make messes because we won't be able to afford the cleanup. The new rules will work toward greater balance between economy and ecology. In working through these rules, we will rediscover the natural world and our place in it. Check those in which you now participate.

New Rules for Business

1. Create no waste
2. Close all possible systems
3. Make the seeds of new creation abide in each product and by-product
4. Design for the long term
5. Work with least effort and greatest efficiency
6. Invest in the energy flow
7. Honor the community of nature from which you draw your resources
8. Spend only the interest, not your natural capital
9. Become a biotic citizen
10. Learn to read the book of nature

Each product must break down into the raw materials for new products, not cubic feet of landfill space. Each dividend must include a portion to natural capital, the planet's energy system. Each corporate citizen must become a citizen of the biota—its bioregion—in which each resides biologically, not just economically and politically. Each must learn to read nature's signals.

These rules are nature centered and rightly so. We can no longer afford to be exclusively human centered. As we shift to being nature centered, something wonderful happens: We find our true and best place on the planet. We realize that something infinitely larger than humanity unites us and that is life itself—abundant, complex expressions of life.

At first nature might seem intricate and foreign. What does it mean "to read the book of nature"? How do we "become biotic citizens"? The very thought is terrifying to many who have insulated themselves from all but a little natural contact. Americans now spend most of their time indoors. What do we know of nature? Only what we visit on weekends or vacation, only what we fight with pesticide and fungicide and herbicide around the house. How, then, shall we begin to make friends with the planet?

Begin by taking a slow walk and looking at the natural fragments you can find in your neighborhood. Go to a park and imagine that if this were an untouched ecosystem, it would be swarming with life. Imagine the array of creatures you would see beyond the occasional squirrel and blue jay. Realize how we have stripped the system of its complexity, have conquered, simplified and controlled the world around us.

Resolve not to add to the process of simplification but to work toward complexity. Work toward restoring the natural landscape, toward restoring some of the natural capital we have so thoughtlessly spent. In my yard I planted a butterfly garden full of plants native to the area. Even a small gesture like this makes me feel closer to nature.

How do you become a biotic citizen? Become aware of the rhythms of the natural world. Look for the cornerstones of life working on this planet and appreciate your increasing awareness as you relish their ceaseless work cropping up in thousands of different ways. As intricate as nature is, the vast power of energy we call the ecosphere really runs very simply: birth, life, death. In all things let us remember the beauty of life on planet earth.

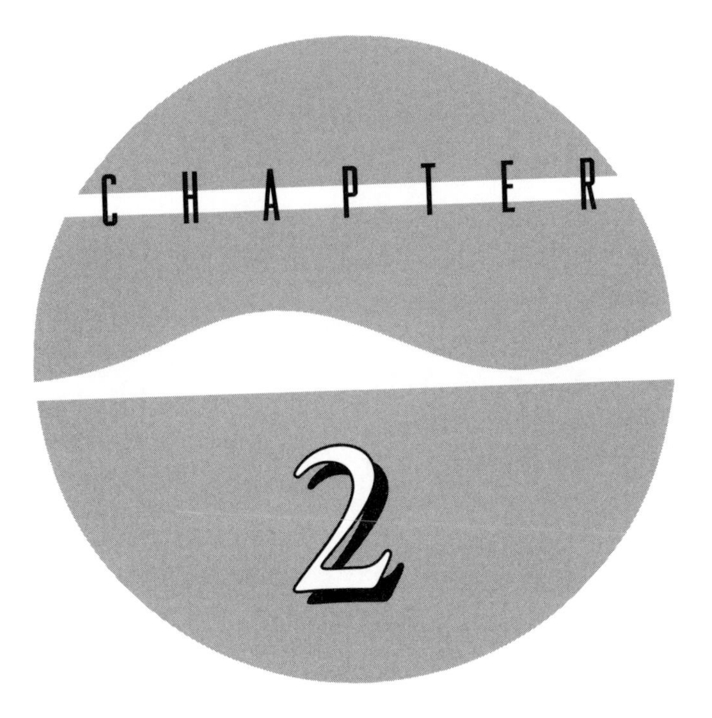

The corporate environmental leader knows where we've been, where we are, and where we're going.

Respect the Tide of History

Why are we where we are in regard to environmental issues? Why all the fuss? Is there really a problem as great as the soap-box arguments would has us believe or is that doomsday rhetoric we ought to ignore?

Perhaps you've pretty much dismissed the claims of environmentalists as exaggerated and emotional, and carry the burden of regulations as inhuman bondage. Maybe your organization is engaged in energy-saving or recycling programs because they make good business sense. These decisions may not have a great deal to do with environmental concerns. A good business decision doesn't have to be tied to a larger issue to make good business sense. Maybe you resent the very whisper of "eco." Yet no person or organization can separate itself from its own times. And, like it or not, we find ourselves at this moment in history in a white water passage. We stand poised to go forward in environmental leadership as the rest of the world watches and applauds our progress. We stand on centuries of hard-won lessons, years of legislation and billions of dollars spent cleaning air and water, protecting species, restoring farm and rangeland—all contributing to the quality of life we enjoy. The time line in this chapter will show where we are and offer some respect for the tide of history and the point at which we find ourselves: Environmental stewardship has broken into the world's consciousness.

At this unique moment in time we have the power to communicate around the globe, instantly to process and forecast information, to travel to remote places and report their condition, and even to view the planet from outer space. As a result of having brought our technology to such a pinnacle, we have also brought our environmental awareness to new heights. We can now see the present state of the entire planet. There isn't a spot anywhere on the surface of the entire sphere that is secret to us. And, we can compare this beautiful, living blue marble to the dead faces of other planets in our solar system and appreciate the difference.

We can now analyze the past and foretell the future. We no longer must repeat the mistakes of previous generations out of ignorance. We can repeat them out of greed, laziness or stupidity, but we can no longer plead ignorance that we didn't know the consequences of our acts.

Over the past five years I have given a lecture on environmental history to many audiences. More than almost any other information I share, this march through our collective history impresses people. As they watch the unfolding of one civilization after another, then see them wither and fade from misuse of resources, the heads begin to nod and the attitudes shift. For the first time, they really get why we are where we are, where we've been and where we're going.

My lecture is distilled into a time line and I present here only the portion pertaining to the United States and its environmental laws. This portion, to which I have added a great level of detail on environmental law, represents only the tip of the iceberg of environmental history. The story really begins in the Fertile Crescent along the Tigris and Euphrates Rivers more than five thousand years ago and continues to the present day in all parts of the globe. Between then and the modern era, enormous damage has been done to the earth in the name of progress. Though voices always rose against such waste and carnage, they always fell into the distant echoes of the past virtually lost to the generations ahead. A hundred and thirty years ago George Perkins Marsh published virtually the first modern observations about the demise of ancient civilizations and the misuse of resources that brought them down. Only in even more recent times have we begun to respond to such knowledge.

One of the great lessons environmental history teaches is the repeated spending as if there is no tomorrow of a nation's natural capital. Not even now are we clear that when a nation spends its natural capital, its future is gone. The citizens have only their past glories as a reminder of the riches they once had but have no more. Nature teaches a cruel lesson. When resources are gone, they're gone. The United States has made remarkable progress in learning this lesson, but we find we must teach and reteach the value of this progress with every change of political power and to each new generation.

The good news is that a good number of people and organizations, perhaps a critical mass in this generation, has awakened. What forces have given rise to our recent consciousness? Why are we collectively such slow learners? History is as complex as our current lives are, but it does deliver up a number of factors that bring us, at last, to our current understanding.

The communication thread

Communication is vital to passing on lessons learned from one generation to the next and from one culture to the next. Until communication speeded up electronically in this century, lessons of the land were slow to be passed on from generation to generation. The ancient Greeks failed to benefit from the failure of their Eastern neighbors to conserve their soil and forests because the lag time from one geographic area to another, from one millennium to another was enormous. The Romans ignored the Greeks' demise from natural causes and then sowed some of the seeds of their own destruction through ruthless development, habitat and species loss, overfarming and grazing. And so it has gone throughout both Europe and the Far East to the present day. Our modern-day ability to communicate visually and verbally around the globe now makes our mistakes more obvious.

The transportation thread

Transportation is another link in our education-delivery system. Viewing places firsthand shows us the necessity for change. More sophisticated transportation also makes it easier to leave a place once its resources are gone. It makes more areas accessible to penetration and requires more fuel to operate. The age-old quest for new frontiers began in Mesopotamia more than 5000 years ago and continues today—powered by the use of and need for new resources.

The energy thread

The search for energy propels humankind forward and around the globe. Energy is the heartbeat of the planet. From the discovery of fire to the wood age, in which a majority of earth's population still live, to the fossil fuel age to the nuclear age, we live and die for a usable energy stream. Now we stand on the threshold of the solar age. As at all such thresholds we have a myriad of choices to make in terms of availability of energy, combinations of energy sources from solar to wind to geothermal and traditional sources such as natural gas, and commitment to new technology that will change the quality of our lives.

The time thread

Now that we have explored the globe, we have met ourselves in the farthest corners. No new continents to conquer appear across a misty ocean. No new planets become available in the night sky. A certain realization sets in on our consciousness: This is home.

The farther back in history, the less technology, the longer a civilization takes to falter and fall. The more modern the civilization, the greater the energy flow through, the greater the destructive technology, the shorter the time span. Every civilization, from the Sumerians, to the Romans, to the present day has risen from pioneering, to luxury and plenty, to fall into relative poverty unless environmental reform and innovative technology or a new source of resources intervene. Our discovery and use of fossil fuel in this century has revived some of these faded economies.

The conservation thread

As the resources thin over time and the pollution thickens, as the natural capital of trees, ore, fisheries, game and wilderness habitat gets spent by each civilization, the citizens wake up and pass more and more punitive laws to reverse the process and try to regain what they have lost. The Greeks did it in the 4th century before the Common Era after the peninsula was denuded. The Venetians did it in the 14th century in the

Common Era. The English did it in the 17th century. The United States began doing it in the early 20th century. Unfortunately, the age-old adage often applies: Pounds and pounds of cure cannot reverse what an ounce of prevention could have prevented. Judging from the number of laws the United States has passed in the past 25 years, we either find ourselves following this learning curve as others before us have, or we are trying to beat it and avoid that kind of hard lesson. Which do you think is the case?

The time line in this chapter begins with the Industrial Revolution in America in the 1800s and brings it to the present time. You will have to imagine all that came before it, the rise and fall of the Mediterranean as power center, the clearing and mining of Europe, the same for the Middle and Far East. The very important knowledge of rise and fall of nations from spending their natural resources had not been appreciated and is rarely taught. Even now the generation who presides over industrial and natural resource decisions in our lifetime has no historical perspective.

This information is by no means complete. You will bring your own set of facts, dates and authors to the time line. However you read it, notice that we seized the resources of the new world with a vengeance. On a continent rich with resources and filling with Europe's poor, thoughtlessly plucking the proffered abundance seemed as appropriate as did annihilating the Native Americans. In Europe, not having enough of the basics was a life-long condition for most of the population. In America, on the other hand, land, trees, game, fish were there for the taking in numbers and quality never experienced in the old country. For the first two hundred years colonists cleaned off the Eastern Seaboard. As the Industrial Revolution got under way, the flow through of energy and resources increased along with the population and spread westward.

For every action there's reaction and the reaction finally came to a head in the late 19th century as people began to realize they were witnessing what had happened in Europe as the continent's last great places were being discovered and defiled daily. Voices in the 19th century built to a crescendo. The Sierra Club was formed in 1892 to protect our treasures. The Conservation Age began and then came the rise of Environmentalism, and now the realization that economy and ecology are two sides of the same coin—in fact, they *always* were.

The sum total of your excursion into this stream of history will show that we have acted on short-term gain, a longing for convenience, ease and comfort, and a love of growth and development absolutely divorced from nature's systems. How we're going to change our ways remains to be seen, but clearly change approaches—enormous changes, dislocations, shocks and pattern shifts will mark the 21st century as we grapple with deep questions about what constitutes the quality of life and who should enjoy it.

History books differ in their reporting of dates and events. Those included here have been verified, but forgive the inaccuracies you may find. The purpose is to show the flow, not the isolated incident.

AN ENVIRONMENTAL TIME LINE

EVENT	DATE	COMMENT
INDUSTRIAL REVOLUTION IN AMERICA	1800s	Increased the flow from argriculture-based to urban-based societies
Englishman Thomas Malthus proposed theory of overpopulation	1803	The first warnings appear of famine due to overpopulation
Cedar waxwings sell for 25 cents a dozen in Philadelphia meat market	1807	Abundant wildlife led to easy hunting and sale of exotic species
Whaling worldwide	1830–60s	290,000 whales taken from the world's waters. Helps grease the growing Industrial Revolution. Remaining whales saved only by discovery of oil in Pennsylvania later in century
America warned by Thoreau, Emerson, others of overuse of resources	1832+	Beginnings of concern over exploitation of American resources
Last elk killed in the Adirondacks	1834	Eastern seaboard cleared of much of wildlife
World population climbed	1850	Reached one billion mark

EVENT	DATE	COMMENT
River boats burned 18–22 cords of wood per day each up and down U.S. waterways	1840s–70s	Increased deforestation of Midwest plowing of the tall grass prairie. Beginning of cattle drives on the Plains
James Audubon painted native bird portraits	1840s	Recorded many new species, over 1000 native bird portraits in all
Chief Seattle's famous speech	1857	"Man did not weave the web of life; he is merely a strand in it. Whatever he does to the web he does to himself"
Origin of Species published in England—the theory of organic evolution through natural selection	1859	Author Charles Darwin connected man to nature, web of life, in a new way
Discovery of oil in Pennsylvania	1859	Starts the modern era based on fossil fuel consumption
Man and Nature published	1864	George Perkins Marsh traced demise of civilizations due to misuse of resources; the message just now being heard in 1990s
German biologist Ernst Haeckel first used the term "oecology"	1866	Denoted the study of organisms and their interaction with their world
Van Buren County, Michigan, sent 7,500,000 passenger pigeons to east	1869	Considered a delicacy for Eastern markets
Golden Spike unites East and West coasts	1869	American locomotives estimated to have burned 19,000 cords of wood per day
Yellowstone National Park established	1872	The first such public reserve in the world

EVENT	DATE	COMMENT
50 million buffalo killed on the American Plains to defeat Native Americans	1872–1883	"Never before in all history were so many large wild animals of one species slain in so short a space of time."—Teddy Roosevelt
AMERICAN CONSERVATION MOVEMENT BEGINS	1872	U.S. government set aside 2 million acres of forest
Oceana County, Michigan, sent over a million passenger pigeons by rail to eastern cities from flocks numbering in the millions	1874	Railroads accelerate the slaughter of birds by making commerce a large-scale operation
U.S. declared the frontier closed	1890	U.S. population is 62,949,714
Forest Reserve Act passed	1891	Assured future supplies of timber. Established the responsibility of the federal government for protecting public lands from resource exploration
FIRST STAGE OF ENVIRONMENTAL ACTIVISM		Natural resources are conserved in small geographic areas for specific solutions at a small, localized cost; resources are seen as supply for human benefit only
Sierra Club founded by John Muir, "the voice of the West"	1892	Sierra Club in 1994 had 500,000+ members in U.S. Their mission to "explore, enjoy and protect the wild places of the earth"
Few hundred buffalo left out of 70 million	1892	Given refuge in Yellowstone National Park
First organized garbage pickup in New York City	1895	Citizens get first real relief from age-old problems of disease, water pollution, trash
Theodore Roosevelt leads conservation movement	1901–09	Conservation's Golden Age

EVENT	DATE	COMMENT
First federal refuge created	1903	For brown pelican off the east coast of Florida
National Audubon Society formed	1905	Purpose to protect wildlife
Gifford Pinchot, first chief of Forest Service, proposed to manage forest resources according to principles of sustainable yields	1905	One of the many milestones of this era
Congress created the National Park System	1912	Law set aside parks to conserve scenery, wildlife and natural and historic objects
Last passenger pigeon died in the Cincinnati Zoo	1914	Brass plaque marks spot of pigeon's demise
Sardine industry takes off in California	1915	Cheap canned fish created Cannery Row in Monterey
National Park Service established	1916	Parks to be maintained in a way that leaves them unimpaired for future generations
PCBs (polychlorinated biphenyls) initiated to stop transformers from exploding		Replaced flammable oils; PCBs discharged into nearby rivers
Animal Ecology published	1927	British Biologist Charles Elton became "father" of ecology, defined concept of niche
CFCs developed	1930	Refrigeration revolutionizes our lives
Dust Bowl in Great Plains	1930s	Overgrazing and farming leads to soil loss during drought years

EVENT	DATE	COMMENT
SECOND WAVE OF RESOURCE CONSERVATION	1930s	Franklin D. Roosevelt attempts to get the country out of Depression by providing 2 million jobs planting trees, providing flood control, decreasing soil erosion, developing parks, etc.
Soil Erosion Service created	1933	Correct massive erosion of the farms of Great Plains states
Taylor Grazing Act	1934	Regulate grazing of domesticated livestock on public lands
Federal Aid in Wildlife Restoration Act	1937	Levied federal tax on all sales of guns and ammunition. Matching funds have provided more than $2.1 billion for states to buy land for wildlife conservation to support wildlife research, reintroduce wildlife
600,000 tons of sardines caught per year off the coast of California	mid-'30s	Cannery Row in Monterey at its height
Federal Food, Drug and Cosmetic Act	1938	Established by FDA to regulate food and drug additives
Sardine fish population collapsed from overharvesting	1940s	Cannery Row collapsed
NUCLEAR AGE BEGINS Atom bomb dropped on Nagasaki and Hiroshima	1945	Scientists claim that the peaceful uses of atomic power will outweigh its immense harm. "Atoms for Peace" program in 1950s promoted nuclear power. as source of electricity

EVENT	DATE	COMMENT
First air pollution disaster in U.S. in Donora, Pennsylvania	1948	6000 people fall ill, 20 died from breathing polluted air
First Federal Water Pollution Control Act	1948	Required states to provide plans for controlling inter-state pollution
Sand County Almanac published	1949	Aldo Leopold eloquently described life on a Wisconsin farm and warns of man's folly with the land; this work still a classic
Population of the world now at 2 billion, 500 million	1950	Population has doubled since 1850
Delaney Clause passed in U.S.	1958	Prohibits sale of foods containing additives that cause cancer in humans or animals
Silent Spring published	1962	Author Rachel Carson warned of the dangers of pesticide use
AGE OF ENVIRONMENTALISM BEGINS Second wave of environmental activism began with the goal of environmental quality	mid-60s	Problems are now widespread, often found in cities; results are long-term, often indirect effects and increasingly harmful to human health. Costs for cleanup have risen proportionately. Focus is still on humans in the ecosystem
High concentrations of air pollutants accumulated in the air above New York City	1963	Killed about 300 people, injured thousands
Clean Air Act passed (renewed 1965, 1970, 1977, 1990)	1963	First national legislation in the U.S. aimed at air pollution control
Wilderness Act	1964	Prohibits development of wilderness areas and established procedures for designation of new protected areas

EVENT	DATE	COMMENT
Foam appears on nation's streams	mid-'60s	Nonbiodegradable laundry substances and cleaning detergents the cause
Lake Erie severely polluted	1960s	Large numbers of fish die, commercial fishing drops sharply, many bathing beaches close
Ecology emerged as a science	1960s	Everything in relationship recognized at last
Man landed on the moon	1969	New view of Earth from space showed how beautiful and fragile is the planet on which we live
Cuyahoga River in Cleveland, Ohio, caught fire from accumulated pollution	1969	Two bridges burned by the five-story-high flames
Oil spill off coast of Santa Barbara, California, caused alarm over off-shore drilling	1969	Increased environmental concerns
National Environmental Policy Act passed in U.S.	1969	Declares a national environmental policy and creates the Council on Environmental Quality; requires Environmental Impact Statements
First Earth Day celebrated	1970	Ecology and environment become popular movement; rise in general citizen concern
Occupational Health and Safety Act passed	1970	Safety and health, on-the-job protection from industrial hazards, hazardous waste maximum levels of exposure, etc.

EVENT	DATE	COMMENT
Clean Water Act passed (renewed in 1972, 1977, 1981, 1987)	1970	Sets standards for waste water treatment, sludge management, establishes effluent limitations and water quality standards
Environmental Protection Agency formed	1970	Nixon administration formed new office out of many old ones—Federal Water Pollution Control Administration, Air Pollution Control Office and others
Federal Insecticide, Fungicide and Rodenticide Act (renewed in 1988)	1972	Registration of all pesticides required, applicant certification, premarket testing
Endangered Species Act passed	1973	Protects endangered or threated species
Chloroflourocarbons are theorize to lower the average concentration of ozone in the stratosphere	1974	CFCs can stay in the atmosphere for years and move up into stratosphere where they attack ozone molecules
Population now at 3 billion+	1975	Third billion reached in 25 years
No new American nuclear power plants ordered since this year	1975	108 previous orders eventually canceled, a result of cost overruns, shoddy construction, cover-ups
Safe Drinking Water Act	1976	Protects nation's drinking water
Resource Conservation and Recovery Act (RCRA)	1976	Regulated the generation, transportation, storage, treatment and disposal of solid and hazardous waste

EVENT	DATE	COMMENT
Federal Land Policy and Management Act	1976	Funded, staffed and enforced grazing policies
Toxic Substances Control Act	1976	Gives EPA custody over assessing risks of chemicals and record keeping
Department of Energy created	1977	President Jimmy Carter attempted to develop long-range, strategic energy plan
U.S., Canada, most Scandinavian countries ban use of CFCs in aerosol cans	1978	Some countries still using CFC-powered spray cans
Three Mile Island nuclear plant goes down. 144,000 people evacuated	1979	Death knell for nuclear power in this country
PCBs banned by EPA		Costly cleanups follow a practice that once solved an acute problem
Gaia published	1979	James Lovelock proposed the earth is a living, self-regulating system
Love Canal residents in Niagara Falls, New York, discover toxic levels of chemicals in their homes and neighborhoods built on old dump sites	1978	Hooker Chemicals dumped 22,000 tons of highly toxic chemical wastes between 1942 and 1955 and covered them with clay
Comprehensive Environmental Response, Compensation, and Liability Act (CERCLA)	1980	Imposed liability on owners, transporters, and generators of hazardous waste; created fund to assist in cleanup costs
Reagan administration relaxed emphasis on environmental enforcement	1980s	Funding for research and development on environmental enforcement sources cut back
California's Cannery Row becomes tourist attraction	1980s	Monterey Aquarium opens; its message is an ecological one

EVENT	DATE	COMMENT
Union Carbide pesticide plant in Bhopal, India, releases toxic methyl isocyanate gas	1984	As many as 20,000 people suffered blindness, sterility, kidney and liver infections, etc. 5000 death toll predicted by 1991. Initiated "Responsible Care Program" for chemical industry
Superfund Amendments Reauthorization Act (SARA) enacted	1986	Required industry to disclose publicly chemicals and toxic hazards in their operations in addition to other CERCLA provisions
Emergency Planning and Community Right-To-Know Act (EPCRA) discloses extent of toxic waste created in pursuit of Consumer Society. Toxic Release Inventory (TRI) now published yearly	1986	Both public and corporations alarmed to learn extent of toxic waste created. Immediate reductions resulted from this airing of bad and wasteful practices
Montreal Protocol gathers countries to phase out CFCs by 1995	1987	Hole over Antarctica the size of United States during winter months discovered
Last California wild condor taken to San Diego Zoo	1987	Captive breeding program increases number. Some returned to the wild
American bald eagle's presence threatened by pesticides, loss of habitat, breakdown of food web	1988	Numbers reduced from 250,000 in 1782 to 11,800 currently
George Bush elected President		Announced he'll be "Environmental President"
THIRD WAVE OF ENVIRONMENTAL ACTIVSIM Understanding shifts to include biosphere, increasingly centered on equality with nature **AGE OF ECOLOGY BEGINS**	1988	Issues are very complicated, involving ecosystems and peoples worldwide. Multiple environmental issues involved in any situation. Responsibility and cost equally complex. Prevention emerges as best choice

EVENT	DATE	COMMENT
Alarm over landfill closings promoted recycling not seen since WWII	1988	Composting and recycling suddenly become household activities. Most cities pass recycling ordinances
DuPont announced it will phase out production of CFCs and halons	1988	International Ozone Trends Panel linked CFCs to depletion of stratospheric ozone
Average global level of carbon dioxide in the atmosphere has increased 25% since 1860	1988	Global warming and greenhouse effect become real factors to contend with; 80% of CO_2 comes from burning fossil fuels; 22% of world's CO_2 contributed by U.S. alone
Rise of anti-environmental groups	1980s	Anti-environmental tactics center on both humans and animals; "wise-use movement" gains strength. Environmental claims in such books as *Population Bomb* and *Global Report 2000* refuted
Brazilian rainforests burning seen nightly on TV	1988–9	*Time* magazine calls this "the year the planet fought back"
Exxon Valdez 11-million gallon oil spill in Alaska rallies new concern	1989	Shock waves roll through U.S. both in communities and business world; still being felt
Amateur films of dolphins caught in tuna nets raised outcry	1990	"Dolphin-free" tuna labels soon appear as customers demand action
EPA's 33/50 Program asked for voluntary pollution prevention		33% reduction by 1992 of releases of 17 highly toxic chemicals with a 50% reduction by 1995

EVENT	DATE	COMMENT
Twentieth anniversary of Earth Day	1990	Outpouring of concern as business and industry take on environmental goals, create relationships with environmental groups; 75–80% of citizens consider themselves to be sympathetic
Pollution Prevention Act	1990	Provides the link between environmental protection and economic productivity
Oil Pollution Act	1990	Required businesses to develop response plans for addressing releases of oil
Corporations rush to create environmental policy statements.	1990	Environmental vice-president often a new office. Public relations on environmental issues become important
Eco-city movement begins	1990	First meeting in California to design sustainable cities takes off around the world
World population at 5.5 billion	1990	Population of United States is 248,709,873. Around the world population increases an estimated 10,000 to 14,000/hour; 1 billion go to sleep hungry nightly. According to UN, 100 million people experience acute shortages of fuelwood. A new, cheap source of energy is needed
Energy Policy Act (EPACT)	1990	10% of new vehicles purchased by state-owned fleets must be alternative fuel vehicles by 1998, private and municipal fleets 20% by 2003
Hazardous Materials Transportation Act	1990	Governs transport of hazardous materials

EVENT	DATE	COMMENT
Business Council for Sustainable Development formed	1991	High-level executives from major international companies band together on behalf of the planet
Green Seal and Green Cross standards for consumer products and processes began to be formulated	1991	Consumer demand for greener products and validation of green claims requires some independent validation
EPA's Green Lights Program launched	1991	Promotes profitable investments in energy-efficient technologies
Rio Conference on Environment and Development held in Brazil	1992	World's nations gather to pledge cooperation on environmental issues. Economy and ecology are seen as two sides of the same coin
First avowed environmental politician makes it to White House	1992	Vice President Al Gore, author of popular *Earth in the Balance*
Large U.S. companies begin to see environmental issues as important market position	1992	"Beyond compliance" attitude becomes popular as companies recognize customers expect them to do their best for the earth and doing so saves money
Environmental entrepreneurship flourishes as new products and services take off; "green" is in	1992	"Benign by design" "waste as resource," "sustainability" become bywords
NAFTA passes amid great controversy	1993	Raises questions of inequities in environmental laws between Mexico and U.S. General Agreement on Tariffs and Trade threatens to lower environmental quality for trade

EVENT	DATE	COMMENT
Electric cars beginning to be advertised	1993	Detroit ambivalent about electric car business
Closing of military bases in United States uncovers toxic waste at sites around the country	1993	Hazardous waste cleanup now a major industry in the U.S.
Clinton adminstration announces Partnership for a New Generation of Vehicles (PNGV)	1993	Big Three automakers and U.S. Government partner to develop automotive and advanced manufacturing technologies
EPA created Pollution Prevention Policy Statement	1993	Provides framework for integrating prevention into all EPA program activities
EPA Energy Star Program initiated	1993	Desktop computers and laser printer manufacturers will introduce equipment that "sleeps" when not in use to reduce use of electricity
Renovated Love Canal homes for sale	1994	How quickly we forget
Environmental Technologies Act	1994	Senate passes bill making companies developing environmentally superior technologies eligible for $236 million in matching grants
Low Income Home Energy Assistance Program passed in Congress	1994	$1.225 billion appropriated to increase energy efficiency
Jury awarded damages of $5 billion in *Exxon Valdez* oil spill settlement.		Exxon plans to appeal
Ozone depletion in stratosphere conclusively linked to CFCs	1994	Volcanic eruptions contribute to only a sixth of ozone loss

EVENT	DATE	COMMENT
25th anniversary of Earth Day in U.S.	1995	More seriously than ever we ask, "If the earth is our home, why do we treat it so badly?"
Contract for America seeks to reverse and loosen environmental regulations	1995	"Takings" movement and "bad science" charges threaten to reverse progress made in years of environmental improvement

We are poised at a critical moment in environmental and human history. As America's environmental progress crests in the 1990s, will the tide recede and much hard-won improvement be reversed? Or will the momentum of centuries of rising awareness carry us on forward into environmental leadership in the 21st century? The next few years will tell.

Former senator from Colorado Tim Wirth wrote in 1992 that "nature is the next great superpower." Actually, it always has been; we simply have not understood that. Undoubtedly, this realization could become our central focus in the 21st century. The progress we make on recognizing and acting on this truth depends to a large degree on the willingness of the corporate world to take the lead in restoring and preserving our natural capital.

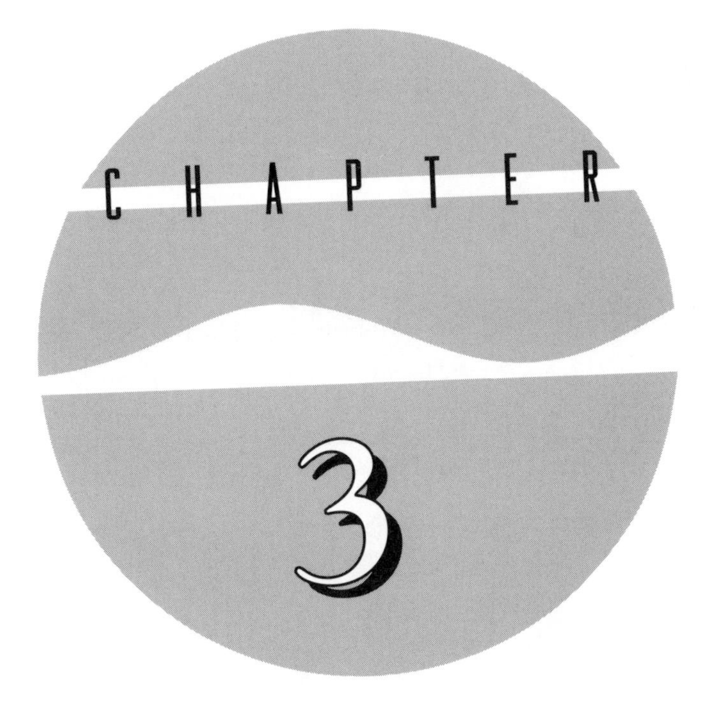

The corporate environmental leader embraces a positive vision of the larger ecology and passes it on to others.

Expand the Vision

A review of environmental history can be a sobering exercise, but we can also find a new direction from it. We should lead the way toward a global economy that honors planetary citizenship. We have the opportunity to formulate an environmental ethic for future generations that will prevent further disintegration of the planet's life systems.

Where do we begin? The world beyond each organization contains both our ecosystem and our econosystem. Both need our help. As environmental leaders, it's our task to connect all the pieces of both spheres into a coherent whole—a heroic but worthy challenge.

The good news is that a positive vision already exists. The following Foresight Principles were written in 1990. We use the word foresight to call attention to the belated foresight in which we now engage. As we look back over human history, it's easy to see we haven't practiced much to this date. However late we begin, foresight is certainly better than hindsight.

The Foresight Principles constitute eight widening circles of influence in our lives. These are written from a personal, not corporate, point of view. Every organization, every nation, is made up of individuals and operates to the level of consciousness and understanding held by those individuals. Each person leads a life that ultimately affects many spheres of influence and activity. The collective impact of these spheres affects the entire planet. The principles attempt to identify those spheres and define ideal ethics and behaviors.

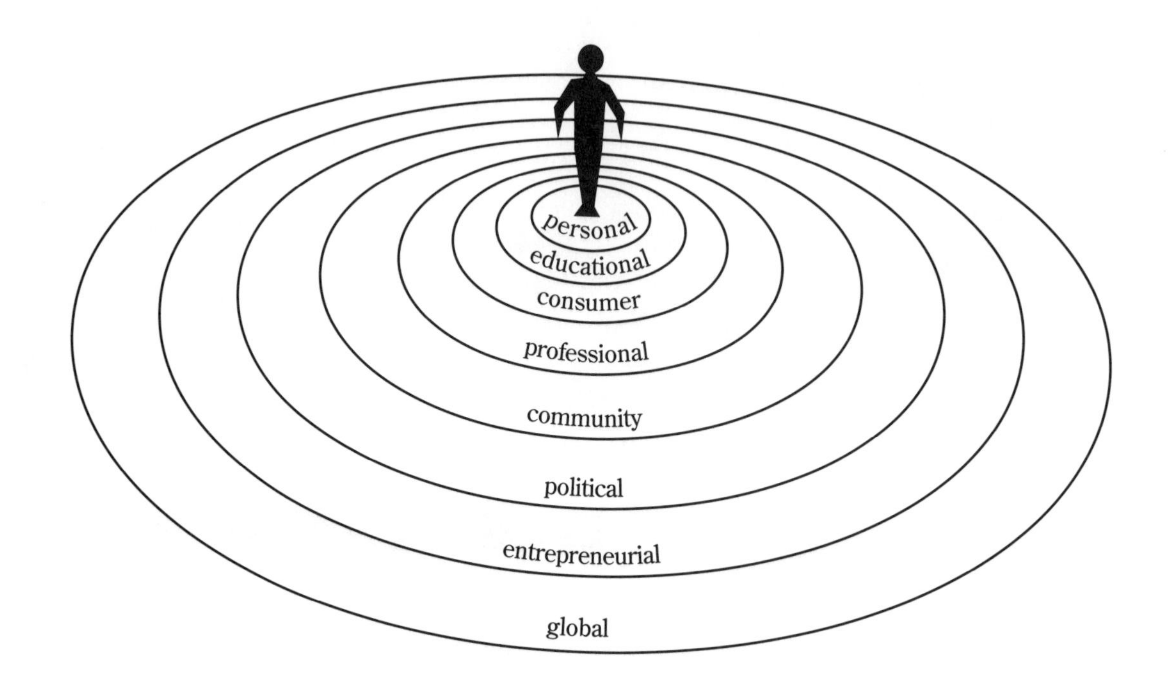

To visualize the Principles, imagine drawing eight circles around you, each one larger than the first and each encompassing a larger area of your life. Stand in the center of these eight concentric circles and walk through them one by one.

The first circle is the home—the place where we really show our values. We have to make changes there before we can tell others to change their ways somewhere else. When we change our own habits, we can sympathize with others as well as guide them to change theirs. Inventory your own shift of habits here. How many have you really made? Have you changed shower heads, started composting, changed light bulbs, started recycling, added insulation? If not, how can you expect the rest of the world to do what you are not willing to do?

In the second circle we educate ourselves. Environmental illiteracy is out. One of the riskiest things we can do is to let others feed us our environmental information. Perhaps the hardest part of learning is deciding what is true among all the sources of information. Clear, concise, accurate information is difficult to come by. At times, nothing less than detective work reveals the simple facts of a situation.

Of course, a firsthand view works best of all. While it isn't possible in every case to go and look, nothing beats an eyewitness account.

Whenever possible, try for a close-up and personal view, even if you have to take your vacation to visit a spot. Eco-tourism is a thriving industry. You can enjoy yourself and learn something at the same time.

The third circle is the consumer circle—what we buy and use and what we do with it afterward. Every dollar is a vote. Any corporation knows how elusive the dollar is and how powerful is the voice of a single consumer call. Since we live in a first-world economy, we consume more than other societies. We must recognize our dollar power and exercise it effectively. If we were to use each dollar as a vote for a new way of life every time we buy something, change would be dramatic. We would buy the products that are most benign and longest lasting. The green market would grow by creating the products that serve us and the earth best.

Right now, we still let convenience and cost dictate our purchases. We want that green future in the abstract, but we want it delivered to us without any effort on our part. Habits die slowly. Yet even small choices at this level reverberate through the halls of corporate giants.

The fourth circle is the professional circle—where we spend our life energy most of the time. We have greater obligation and greater opportunity here than perhaps anywhere else. Here our influence can have great effect, but here our words can fall on ears deadened by familiarity. Small changes here make huge differences in energy, waste reduction and process change. Changes within the walls of an organization make big results everywhere.

The fifth circle of influence widens to the community—our immediate econosphere resting on our immediate ecosphere. In the community both "ecos" meet in close and gritty connection. Environmental justice and racial equality go hand in hand. We live and work here, breathe the air and drink the water Congress has passed laws to protect.

The sixth circle recognizes politics as a vital expression of the public will. The politics of the environment attempt to save us from irreversible pollution. Legislation brought about because of public concern is designed to protect us from the fouling of our natural resources.

The seventh circle of influence constitutes environmental entrepreneurship. While large companies save money reducing waste, small, new companies make money using waste. A symbiotic relationship grows. Environmental technology will be the leading edge of innovation in the next decades. Congress, recognizing this, has appropriated money for such innovation.

The final circle encompasses the living planetary systems and all earth's creatures, great and small, human, plant and animal. At last we celebrate the entire globe and our mutual dependency. We know that we affect those on the other side of the globe and they affect us. We need to know more about the global patterns of the econosphere-ecosphere mix of systems, so that we can respond with an environmental ethic that works for all parties.

Now that you have visualized the eight circles around you, the following principles will illustrate the areas of influence they represent. Each describes a new way of living that leads to personal and planetary stewardship.

The Foresight Principles of Environmental Leadership

The Declaration of Dependence on a Restored Earth

We the people of this planet earth, realizing the unique and wonderful gift of diverse and fragile life we have inherited, do propose and agree to the following principles of foresight for restoring, protecting, conserving and enhancing the quality of all life with whom we share this globe.

1. *A Personal Vow: We Put Our Own House In Order*

We as human beings represent the highest art form of life on this planet. As such we have the sacred obligation in our personal lives to be most careful stewards of the web of life of which we are its most exotic expression. We know we are connected to all living things in ways we do not yet understand and that even the earth itself is alive and furnishes us with bountiful riches without question. As we accept these gifts we agree to use them with care, restraint and reverence for all life forms and life systems. In return for these bountiful riches, we commit ourselves to taking responsibility in our homes to begin restoring the water, air, forests and fields of this planet to prime health as flourishing ecosystems. We will each begin the change in our hearts as we reconnect with the natural world and in our homes as we center ourselves around this new respect for a healthy planet.

2. *Education Is the Key to Right Action: We Shed Light on Natural Consequences Rather Than the Behavior Itself.*

We agree that education is the key to understanding and right action. We hereby commit to educating ourselves and others in an open, honest, friendly and thorough manner. The natural consequences of continued indifference to the natural world are dire. Clean, truthful information is vital to address the particular challenges each of us faces at the local level. We understand that education is a part of the life of an environmental leader. Communicating that knowledge to motivate others to right action is critical to restoring the earth.

3. *In Consumerism Lies Our Greatest Power: We Require Ourselves To Make an Enormous Number of Simple Choices.*

As citizens we have the power of choice to vote with every coin and must, based on sound information, make an enormous number of simple choices that will slow down use of resources, reduce, avoid and clean up waste, lengthen product life and protect ecosystems, including our own health and lives.

4. *In Our Professional Lives We See Things Backwards, Inside Out, and Upside Down in Search of Eco-Excellence.*

We choose not to be pushed into environmental cleanup by an angry public. We choose instead to lead in environmental change. We will abandon reasons for "business as usual" and look everywhere for ways to change—to lead! As the active support system we call economics provides us with work, products and services, we agree to provide it with a sustained resource base. We will extract with the greatest care only the necessary raw materials, reduce our waste to a minimum, dispose of it in a careful manner for both our health and that of the planet, enhance our use of present resources and return to the earth a replacement of those resources wherever possible. Thus we will create a system of "eco-nomics" that encompasses both natural and human-made systems.

5. *In Our Communities We Insist on Equality.*

As community members, we know that the poor and hungry, the dislocated, ignored, violated and violent members of our human life system must have care, attention, education and lasting solutions. Our human ecosystem is at least as important as our natural surroundings. Failure to provide these services only puts unbearable weight on the lives of those who surround the unfortunate among us. Not only is equality of resources fair and just, it is uncommon foresight in avoiding even greater problems in the future. In this arena alone heroic leadership is at work, must be appreciated, supported and joined by an army of thousands more until we have reached a state of peace and justice for all.

6. *In the Political Arena We Empower Through Cooperation.*

In the course of human events we have at last created the political arena, the rights and the means for free expression and equal justice. We commit to fair access of resources to all peoples and equal representation of all life forms under the law in both war and peace. Rather than agreeing to the lowest common denominator of environmental standards, we commit instead to the highest, to creating a national and world reverence for the health of planet earth and working toward that through global cooperation. We commit to truthful communication and community building. We will *be* the change we want to make. We will vote for those who respect the earth and against those who do not. We will rise above the short-term gain, greed and hate to create a generation of great environmental leaders in every avenue of political power.

7. *In the Entrepreneurial Spirit We Create Opportunity Out of Obligation.*

We know the earth cries out for relief from assault on herself and her creatures—including humankind—and commit to creating the kinds of products and services that will protect and enhance the quality of life on this planet rather than further erode it. Perhaps nothing more than the creativity of the entrepreneurial spirit combined with the best of environmental leadership can produce the change in direction we so sorely need to take. As environmental leaders, we commit to making those changes.

8. *In the Global Arena We Act on Behalf of the Whole.*

We search through political and economic solutions, through cooperation and commitment to a world community forged in friendship and mutual dependence on a restored earth. Though we are individuals collected into individual nations, we honor the whole as much as our part. We understand that what we do on this side of the globe affects those on the other. What they do over there, affects each of us here. Without the understanding of the wholeness of our lives on a global scale, we will continue the mistakes of the past and overpower all hope for the future. We here dedicate ourselves to thoughtful action that anticipates and provides for a healthy, restored world for all living things and those to come.

To these principles we pledge the quality of our lives, the remaining plenitude of this beautiful earth and our sacred obligation as the pivotal generation in the future of life on this planet.

This is a world view not to be chosen lightly. This view means a thousand adjustments great and small to bring the new vision into focus. Almost like turning a camera lens to get a clearer picture, we can see how the world could really work. We're not talking sacrifice, we're talking adjustments. Look at each principle again. What would it take for this change to come about? Mostly just an attitude shift and some changes along the system to slow it down, close it and simplify it. Many small adjustments will relieve the stress of a system overburdened with unsolved problems.

Changing the System

All systems go through four predictable stages, according to Jeremy Rifkin in Entropy: Into the Greenhouse World. They're called stages of entropy in which the flow-through of energy flows faster and faster until the system collapses. They overlap each other as one dissipates and the next emerges from the previous. Check the phase in which you find your system.

Entropy I

The pioneer stage. All energy is focused on getting systems in place, making things work, a period of high innovation during which citizens or employees will willingly sacrifice for the common good.

Entropy II

Systems up and working. Elaboration of systems begins. Refinements add to costs and sophistication. More and more people involved in receiving benefit and in maintaining the infrastructure.

Entropy III

Long-maintained system begins to break down and needs repair. More rules and regulations passed to keep systems working. More complications and elaborations are prescribed to hold the system together. People sense that this system no longer works, but do not yet have strong new candidates to take its place.

Entropy IV

System collapses. People suffer great privation as accustomed system fails them. Chaos may reign. Since it becomes evident the system cannot be repaired or rejuvenated, all energy focuses on regrouping and gathering resources to create new systems—and back we go to Entropy I.

Do any of these stages seem familiar to our lives today? Entropy III kicks in with punitive laws, restrictive checkpoints, mounds of paperwork and elaborate repairs, even while more people voice their dissatisfaction or outright refusal to participate in the system. Life is manipulated by large organizations over which we have no control. Does this ring a bell?

Look anywhere to find evidence of Entropy III. How about the web of environmental regulations we have to obey? They show a response to the complexity of a system that has begun to fail. What was benign yesterday will be toxic tomorrow, creating more regulations to follow or transgress. What we need to do is to quit spending money to shore up the old system and spend our energy building the new one as fast as we can.

This chapter expands the corporate vision beyond the walls of the business to the larger ecology. This larger ecology is at the same time more personal, offering many opportunities to lead in positive changes in the systems that make up our lives.

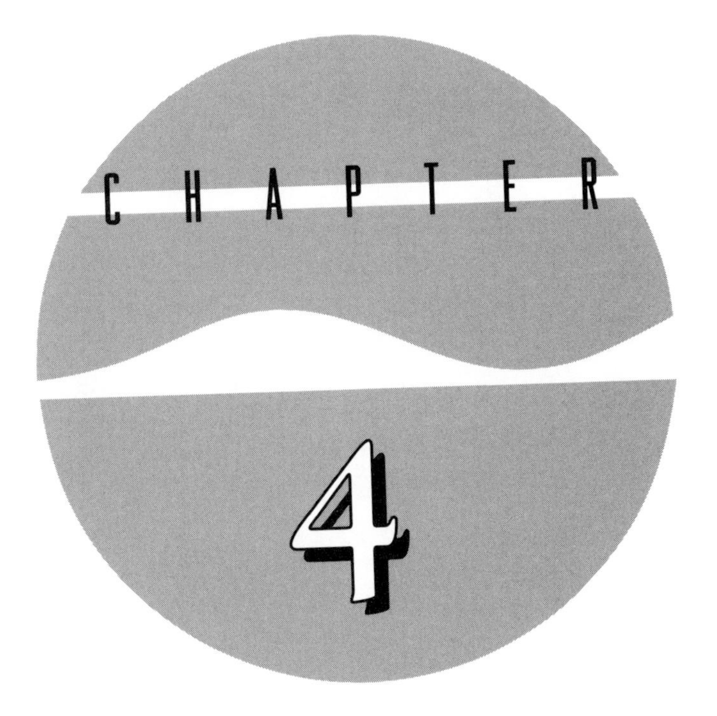

The corporate environmental leader designs a process that emulates nature rather than destroys it.

EMULATE THE PRINCIPLES

Until the passage of EPCRA, the Toxic Release Inventory and the Public Right to Know Act of 1986, industry responded to environmental cleanup and waste control as a daunting, aggravating, thankless, expensive task forced on business by a villainous government reacting to bad science and an emotional public and focused on impeding the progress and profit of private enterprise. Then a funny thing happened.

Companies realized that they could eliminate waste and that could save money—sometimes lots of money—and it was sometimes as easy as making a decision to do so. Not necessarily cheap, but certainly as simple as that. They could fold waste reduction into total quality management as a natural next step. A few companies across the United States have been doing this for 20 years, but most are new to waste reduction and recycling. We the public can now expect to see even more changes in the years ahead.

Now that pollution is global, with enormous social and economic costs, it becomes obvious that help is needed from all parties, certainly from those whose processes have most created it. Forward-thinking companies are jumping on the "beyond-compliance" wagon. Soon those companies that are still polluting will be viewed as hopelessly outdated and noncompetitive, if not insensitive and calloused as well.

Is respect for nature the reason for such change or can companies save and make money by doing good for the environment? The answer to both questions is yes. Respect for the natural world is a good reason for change, but it has to make good business sense for companies to make it work.

In the year of the 25th anniversary of Earth Day, an inventory of corporate effort would seem to be in order. We may ask these questions of our industrial powers: What kind of progress have you made toward eliminating waste? Have you saved money? Made money? How many of your processes have changed as a result of increased commitment to both environmental laws and public consciousness? This chapter will answer these questions.

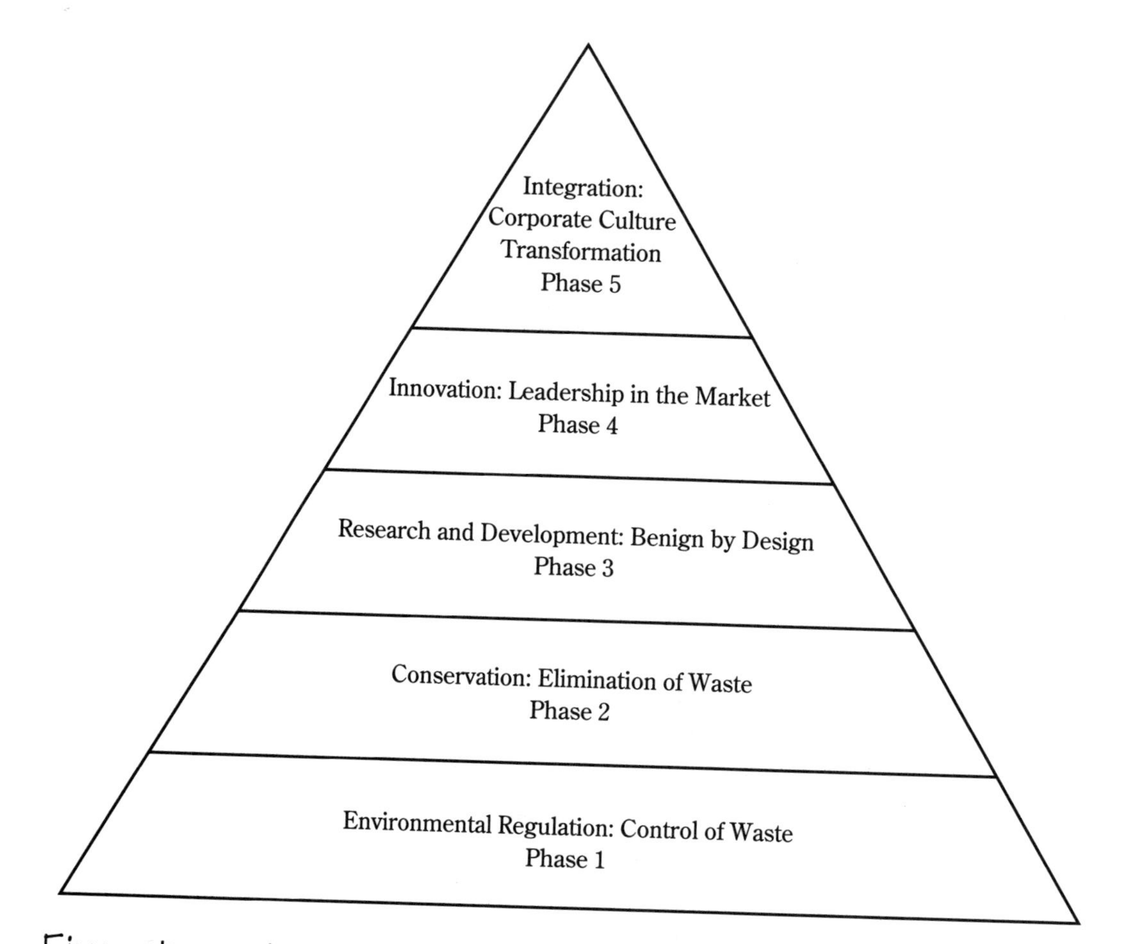

Five steps to corporate environmental leadership

As companies change they go through five phases in their evolution toward a new environmental ethic. Check where your company may be at this point.

☐ 1. *The Regulation Step: Control of Waste*

- Compliance issues in air, water pollution, solid and hazardous waste disposal are the focus.
- Site remediation demands attention.

☐ 2. *The Conservation Step: Elimination of Waste*

- Obvious waste is pinpointed and eliminated.
- Flow-through systems are closed or obsolete and inefficient systems shut down.
- Throwaways are cashed in both by avoiding tipping fees and by reusing materials.
- Corporate culture buzzes with new words, new team approaches.
- Results can be dramatic: 75 to 90 percent reductions in one to five years.

☐ 3. *The Research and Development Step: Benign by Design*

- Teams mature; corporate environmental policies are written or enforced.
- Innovative solutions to pollution problems are researched.
- Prototype products and processes are tested.
- Benign-by-design thinking takes hold.
- Eliminating the last 10 percent of waste poses a real challenge.
- Employee reward programs spur progress.

☐ 4. *The Innovation Step: Leadership in the Market*

- Green products go to market.
- Green processes work and both save and make money.

- The company takes leadership through business and community outreach as well as in production.
- Company leaders realize public image and larger constituencies are a nice by-product of this effort.

☐ **5.** *The Integration Step: Corporate Culture Transformation*

- Old ways are lost in unexpected ways.
- The corporate culture assimilates the new environmental ethic and adjusts itself to it.
- The company undergoes a long and continuous recycling of these phases until each inefficient, polluting practice dies out and is replaced by new approaches.
- Employees are deeply trained and committed to this ethic.

Companies can be engaged in several of these steps simultaneously on different products and processes, which you will see as you read their profiles. The point of these profiles is to mark the shift in corporate awareness from controlling waste to eliminating it, and from old products to new. Ten years ago, even five years ago in many cases, companies didn't have corporate environmental policies. They didn't have offices dedicated to environmental operations and public relations, although they may have been complying with regulations. The larger focus of answering to the public, respecting the planet's life systems and educating and empowering employees to go beyond compliance issues simply didn't exist. It does now.

This is a new concept for corporations. Some changes have been rudimentary, obvious and long overdue. Others are creative and worthy of national attention. I've listed a wide spectrum of innovations to avoid redundancy and to increase inspiration and information for the reader. Almost every company listed is doing far more than I had room to acknowledge. As you read these reports, look for these points.

1. The date from which most of this activity sprang was the passage of EPCRA, the Public-Right-to-Know Act and the Toxic Release Inventory (TRI). Even as the first TRIs were published, news reports showed how

much the earth was suffering from this onslaught, how great was the pollution and how high the cost to the natural systems. This kind of public information prompted public relations and eventually an attempt to show how much each company had reduced waste.

2. The change toward more benign practices imitates natural systems.

3. The full stories of these companies are in the Appendix. The following highlights will invite you to read the whole story of each company, although we have only skimmed the surface of any given organization.

4. Most corporations approached for this report were cooperative and eager to be involved. Some were not. A few were emphatically not. Feelings run strong on both sides of the corporate environmental question.

5. These high-profile companies have products and brand names that are instantly recognizable. All three U.S. automakers are included for comparison, but German and Japanese automakers, whose recycling and alternative fuels programs are more innovative than ours, are not.

Several chemical companies were included to show their comparative strides in reducing toxic emissions. Three retailers—JC Penney, Target and Wal-Mart—are included, as well as some smaller companies. They are a few of the many companies that form the background structure to our life style.

Examples of the Conservation Step: Elimination of Waste

All the companies are at work in the conservation phase. One of the most mature programs in this area is that of 3M Company, whose "Pollution Prevention Pays" program has been in place since the 1970s and has saved the company more than $710 million dollars since then. Other examples include:

- Anheuser Busch reduced the diameter of its beer can lids in 1992 by $^1/_8$ inch and reduced aluminum consumption by more than 20 million pounds per year.

- Hoescht Marion Roussel eliminated the cotton from its prescription products in the United States for a savings of 110,500 pounds of cotton waste, or 1,418 miles of cottonwads per year.

- Target uses a 100% paperless system on domestic purchase orders. This involves 2.7 million transactions and keeps 40 tons of paper per year out of the landfills.

- American Express has installed an optical scanning system for document storage. Each disc stores 500,000 pages.

- Bell Atlantic published 34 million phone books in 1994 using 51,000 tons of paper containing a minimum of 24 percent postconsumer content. This saves more than 10,000 tons of virgin paper and creates 10,000 tons of postconsumer waste which fuels the recycling process.

- The Coors brewery in Golden reduced ammonia releases by 86% in 1993.

- DuPont has a carpet-reclamation program that has diverted more than two million pounds of used carpet from landfills to date.

- At Ford Motor Co. plastic wrap from assembly plants is made into protective seat covers, which saves more than 400,000 pounds of plastic from landfills each year.

- At General Mills a reduction of weight of Hamburger Helper cartons reduced packaging by two million pounds of material per year.

- Engineers at Monsanto's LaSalle plant near Montreal, Canada, eliminated 90 percent of the xylene and ammonia formerly discharged to wastewater by capturing and reusing the liquid waste in closed-loop systems.

Examples of the Research and Development Step: Benign by Design

- Dow Plastic received patents for its 100 percent carbon dioxide blowing agent technology, replacing CFCs and HCFCs in the production of polystyrene foam sheet food packaging. This process makes bowls, cups, egg cartons, meat and cafeteria trays with no harm to the ozone layer. Dow has licensed the technology to six other companies. The full conversion of these licensees will eliminate the use of more than three million pounds of HCFCs per year.

- A partnership among General Electric, Digital Equipment Corp., Nailite, a manufacturer of building materials, and McDonald's connects these companies to recycle computer housing made of GE plastics into roof tiles for McDonald's restaurants.

- McDonald's purchases more than $200 million worth of recycled products annually from vendors at an average of $22,000 per store since 1990.

- Wal-Mart has created an environmental demonstration store in Lawrence, Kansas, which is equipped with latest lighting and energy-efficient systems and made with recycled materials. The store accepts postpurchase packaging before the customer leaves the store and offers a full recycling drop-off site to the community.

- Employee motivation programs abound: Coors has a Scrap (Save, Conserve, Recycle and Profit). Chemical emissions reportable under EPCRA have fallen 57 percent since 1989 as a result of employee-initiated pollution-prevention programs.

- General Motors has We Care (Waste Elimination and Cost Awareness Rewards Everyone), which has reduced solid waste on production lines.

- At General Mills 98 percent of the cartons produced are made of recycled material with at least 35 percent from post-consumer materials.

Examples of the Innovation Step: Leadership in the Market

- Xerox has committed to "Design for the Environment," which means that nothing entering the plant should go to the landfill. Everything should go into production or recycling.

- Gillette has created a "green cell" battery, a rechargeable cell in electric shavers. The energy source is hydrogen stored in a special alloy in the negative electrode.

- Proctor and Gamble has created refill packages for more than 57 brands in 22 countries. Conventional packaging typically costs and more and adds up to 70 percent more solid waste.

- Chrysler offered its Dodge Caravan Electric and Plymouth Voyager Electric in 1994.

- Coca-Cola works with Keep America Beautiful to improve waste-handling practices in communities across the United States.

- Dow initiated a polystyrene cup recycling effort at Chicago's Comiskey Park. White Sox fans recycled more than two million cups in 1993. Dow has partnered with the National Park System since 1989 to recycle more than 900 tons in seven national parks.

- DuPont has expanded polyester recycling capabilities. It plans to convert postconsumer and postindustrial materials once buried in landfills to produce virgin-quality raw materials without the oil-derivative feed stock necessary for new polyester production.

- Ford has established Wildlife at Work, a program of 300 sites worldwide that demonstrates sustainable development and compatibility of industry and the environment.

- H. B. Fuller's Cameo (Computer-Aided Management of Emergency Operations) database inventories buildings with a diagram of each site. The program also models the accidental release of volatile chemicals. Each facility has a spill simulation and worst-case scenario.

- Good Humor-Breyers, in partnership with Boston Edison, retrofitted the heating, ventilating, air-conditioning and lighting in the Framingham, Massachusetts, plant, resulting in more than 12 million kilowatt hours (kwh) of electricity saved for 112 percent of targeted savings.

- McDonald's publishes *WEcology* for 6- to 10-year-olds. It also sponsors the Tropical Science Center in Costa Rica, which works to conserve some of Costa Rica's rainforest.

- Bayer sponsored an annual Ohio River cleanup in 1991. More than 13,000 tons of garbage along 2,000 miles of shoreline was recovered. A 55-acre lagoon, home to alligators, fish, birds and other animals is maintained at its Orange, Texas, plant.

- Target has returned 5 percent of pretax dollars to the stores' communities for environmental, arts and social action activities since 1962. Target and Kodak raised $300,000 for national park preservation and promoted their *This Land is Your Land* campaign.

Examples of the Integration Step: Corporate Culture Transformation

- Xerox has adopted life-cycle analysis principles for their products since 1992, meaning each product's total cost, including energy, is mapped from harvesting raw materials to final disposal. Its "Design for the Environment" includes choosing nontoxic materials and designing for disassembly.

- American Express circulates a questionnaire to employees on their views of environmental program needs within the company.

- Coca-Cola is one of the first major companies to develop a comprehensive environmental education program for employees. Its Environmental Development Program helps employees understand the issues so they can conduct business according to the company's principles and do their part to protect and enhance the quality of the environment. The program has become a model for other companies.

What this represents

Small changes make big differences. Because we produce and consume in such enormous quantities, a $^1/_8$" change on a product can save millions of pounds in resource use and disposal.

Environmental laws bring us closer to a sustainable way of living. We wouldn't be this far without them. The necessity of obeying them brings invention to new levels and adaptation to new heights.

Voluntary programs create positive reinforcement. The EPA's 33/50, Green Lights and Energy Star programs reinforce the positive and increase collaboration and dialog with business. The EPA's shift from being an enforcer to a motivator is welcomed and long overdue. Some companies do the same work without joining any of these programs, while others enjoy the mutual support.

We have a long way to go. These success stories show us only the beginnings, as encouraging as they are. If one company can still emit millions of pounds of pollution into the atmosphere each year, imagine what thousands of companies do together.

We face a long-term challenge. It won't be easy to design and think for the future. Technical problems and social resistance will make a shift to the environmentally sound more difficult.

Consumers must share the blame for the level of waste in which we find ourselves. Look around any American home—it is filled with all the good things of the Consumer Society. We can hardly enjoy the products and blame the producer. As we have been partners in waste building, we now must become partners in waste elimination.

A generation of environmental entrepreneurs is springing up around these corporate giants. As these businesses grow in strength and competitiveness, they will further pressure older businesses to respond to environmental change.

What does our future look like? Imagine, if you can, how corporate America will operate in 2100. We will have banished waste. Paperless offices will be the norm. Replaceable parts and long-lasting, nontoxic, recycled and recyclable products will be standard. Alternative fuels and energy will have replaced fossil fuels. Corporations will be deeply involved in supporting both environmental and social action in their communities. Does this sound too good to be true? Maybe not. The future is what we make it.

The corporate environmental leader overcomes the hurdles to a green future.

Neutralize Negatives with Positives

Moving from compliance to regulations to waste reduction has its ups and downs. It isn't easy to abandon a mature working system that has served the business well and adopt a process that will slow production and cost money. Yet organizations entering this brave new world experience many positives and some unexpected benefits as well. The following list shows some pros and cons of improving environmental efforts.

The Negatives

1. New unfamiliar materials
2. New unfamiliar processes
3. New unfamiliar vendors
4. New unfamiliar skills
5. New unfamiliar technology
6. New unfamiliar partnerships
7. New unfamiliar vocabulary
8. Experimentation
9. Testing
10. Retesting

The Positives

1. Nonhazardous materials
2. Reduced EPA paperwork
3. Savings in waste-handling fees
4. Savings in purchase of materials
5. Better health for workers
6. New partnerships with vendors
7. Team building among employees
8. New partnerships with community
9. New skills for workers
10. People know they are doing the right thing

Introducing the variables

When Hallmark Cards changed its printing processes to use fewer solvents and more water-based inks, a domino effect of changes occurred.

"Every time you change something, you add variables," says Allan Marcotte, manager of Screen Print and Gravure Operations at the Kansas City manufacturing facility. "The more variables you have, the more chances you have to fail."

The conversion to water-based inks in the screen printing operation required almost two years of testing before the end product met Hallmark's quality standards.

"You have to want to use these inks," Marcotte says. The water-based inks initially took the consistency out of press runs, and spoilage and other related press problems increased. To control these issues, Marcotte's staff had to control viscosity and pH more, and learn a new method of making and handling screens. Today the printing quality is as good as ever, Marcotte says.

Steve Walker, lithography manager, worked with many vendors before finding those who could reduce the volatile organic compound (VOC) content of solvents and provide alcohol substitutes to meet Hallmark's needs in lithography printing. He says both Hallmark and its vendors became frustrated in the process; making changes of this magnitude requires a high level of trust among vendors and their customer firms. Even with these challenges, dramatic reductions have been made in lowering VOC solvents and the concentration of solvents in the air. The overall health and safety of the lithography area has been greatly improved.

Introducing recycled paper provided another challenge: In many cases, the quality of recycled paper wasn't consistent and this created another factor the printing staffs had to manage. Every variable had to be introduced, monitored and established in the entire process.

Positive results

What is the payoff for Hallmark?

- The printing process is much less hazardous.

- The volume and, therefore, the cost of shipping and disposing of hazardous waste has been reduced 80 percent.

- Workers learned attention to detail; press operators broadened their experience and technical expertise.

Building partnerships: bottom up and top down

New partnerships, both internal and external, will form around reducing waste and eliminating pollution. Whatever groups they represent, the new networks share information to solve problems.

Larry Long, director of Safety and Environmental Initiatives at Anheuser-Busch (A-B) helped to organize teams in each of A-B's 13 breweries. The teams represented a cross section of the business from top to bottom and across departments. Each brewery attacked the critical solid waste issue beginning in 1990. Teams prioritized items of importance and worked from there. One team videotaped the contents of dumpsters and could tell "shift by shift what went in like rings on a tree." The team attacked the waste until the waste-management company said nothing of value was left to recycle. Workers reduced A-B's solid waste by 75 percent.

A-B also realized that the strapping that holds beer cans on pallets could be recycled. The strapping came in different colors and types of plastics. Vendors and A-B staff worked to change the corporate purchasing specs so that the straps became uniform, and then set up a program to recycle them. The change now saves a thousand tons of strapping a year from the landfill.

Does every employee want to be involved? According to Joe Verga, manager of recycling and waste reduction at Bell Atlantic in New Jersey, age makes a difference: Older employees get "very excited about recycling" because they grew up in the Depression and understand the value and advantages of reducing waste. The middle-aged group is "used to the throw-away society and finds it difficult to change habits. The young ones understand recycling and waste reduction as a natural way of life because they got it in school."

Verga finds that once results are felt and the effort of reducing waste is seen as finding environmental solutions and enhancing the community, most people want to participate. Once they get engaged in the concept, "most people's ideas are not silly," he says. Plus, he says, the program demonstrates our responsibility to be "better corporate citizens in the communities we serve."

At DuPont in Wilmington, Delaware, G. Irving Lipp, senior programs manager, reports the company's new safety commitment replaced the former corporate policy. The new policy was circulated throughout the company and received thousands of comments, passed through internal environmental groups and was signed by the chairman and other leadership. This approach reflects the company's philosophy: No one person has the responsibility; everyone does.

Can one person make a difference? Yes, according to Jack Lonsinger, director of Regulatory and Community Environmental Affairs at Bayer, Kansas City, which produces agricultural chemicals. The assembly line worker is heard more now. "The concept of total quality management has made it more possible for people to speak up. If one person has the ability to bring other people who are strong decision makers together," Lonsinger says, "then you have the possibility [that things can change.]"

Such leadership and support begin at the top. The forces for change at Hallmark, according to environmental administrator Don Martin, were "regulatory driven, competitive advantage, better working environment for employees, cost savings, but more a corporate desire to do the right thing."

"All the benevolent concepts of corporate life—safety, working conditions, environment—have to be top-down driven," says Lonsinger at Miles Corp. "If you have the concept from the top that these things are important, then you get what you want eventually."

Neutralizing the negatives

The disruption of processes, time lost, new skills needed and new technology learned can all be handled. Any negatives within the plant can be overcome if true leadership is available and partnerships function across lines and departments. As Allan Marcotte at Hallmark says, "You have to include the folks. Without their help, it won't happen."

Is the corporation truly turning green or just looking for cost-cutting measures? Al Baker, director of environmental health and safety at Hoechst Marion Roussel (HMR) in Kansas City, believes "everyone has greened quite a bit." He says that most people in the company want to do what they can to improve the environment and that most problems of producing with less waste are "no more than hurdles, not necessarily obstacles." For instance, the removal of the cotton from pharmaceutical bottles was a change requiring some rethinking and refining of HMR process in tablet creation and packaging requirements. "Now that the public is more educated and more packaging sensitive," Baker says, it sees some waste as ridiculous, which makes it easier for the company to be more innovative.

Greater obstacles than internal ones can affect decisions. Joe Verga from Bell Atlantic lists external challenges: the economy, worldwide competition, profits, fluctuating markets and resulting corporate uncertainty. For example, attempting to recycle telephone books is an obstacle course. "Sometimes they want your paper and sometimes they don't," Verga says. "It would be a mistake for any company to think that it can create some rules and walk away with a finished program. It will flounder. Programs must be nurtured." Verga says environmental efforts must be adjusted constantly to meet the demands of a shifting marketplace.

The bottom line is still financial. Until you solve the waste problem, either hazardous or solid, "you throw away dollars and pay more dollars to have it hauled away," says Martin from Hallmark.

The future: new vocabulary, new benchmarks

Corporate environmental leaders speak a 1990s vocabulary that barely existed a few years ago. Some of the new buzzwords sound almost as much like oxymorons as corporate environmental leadership did just a short time back: product stewardship, sustainable manufacturing, tradable permits, lightweighting, packaging sensitive, zero waste, ecologically sustainable production, total quality environmental management, community outreach and right to know, and life cycle assessment or analysis are the words that reflect our direction toward a green future. They describe the ecologically sustainable production created with environmentally sound technology that will become the market edge in the very near future.

What do these corporate environmental leaders see in the years ahead? What are their predictions? Can the last 10 to 20 percent of waste be reduced to zero? Is "zero waste" possible?

"It's like trying to get the last toothpaste out of the tube," Jack Lonsinger from Bayer says. "I can improve my yield on a process. Maybe I can get up to 96 to 97 percent. Getting the last three percent is hard, and to do it the process must be changed. We'll have to go back to square one and find out if doing it in a different way will get you where you want to be."

At HMR, Al Baker says they are beginning to do life cycle analysis on new products, "getting it right the first time—a big change from the past." Joe Verga at Bell Atlantic says they are "looking into products made with recycled materials." Many products Bell Atlantic buys today are imported, and Verga thinks companies in the United States may have to buy some originals but can remanufacture more than they buy here, such as cartridges for laser printers. His point of view is to "look to buy products that are made with recycled materials and products that can be recycled."

While some changes, such as Verga's, are simple alternatives to doing business, others require high technology and new science. At DuPont, which is phasing out CFCs, Irv Lipp says, "We did not change until we understood the science," a change he described as a "task of cooperation." Though changes don't necessarily have to be based on science, new technology will certainly be the key. As Jack Lonsinger at Bayer says, "We will solve our problems with more technology, not less."

The moment of synergy

Although corporate America is still largely in the first and second phases of corporate environmental leadership (regulation: control of waste and conservation: elimination of waste), the next phases (research and development, innovation and integration) can't be far behind. Clearly, 21st-century business will center around bringing the two ecos, economy and ecology into balance.

Environmental managers need three vital components to proceed in their efforts: information, incentives and desire. Those three conditions precede any real change. Once put into place, new processes show significant differences. Managers who work toward a green future find:

- They save valuable materials, raw materials formerly seen as pollution or waste and not sold as product.
- They cut waste-management costs, sometimes deeply, always significantly.
- They reduce their liability under EPA and OSHA regulations.
- They increase teamwork, problem solving and morale. Productivity is sometimes increased because workers are exposed to fewer pollutants, feel better and respond accordingly.
- They promote more efficient allocation of corporate resources through better budget analysis and practice.
- They extend total quality management to include environmental responsibility.
- They learn to choose their "best shot" at the processes that will produce the greatest reduction in cost with the greatest ratio to benefits.

Clearly, the benefits of reducing and eliminating pollution and waste outweigh the costs in many cases, not all of which can be measured on the bottom line. Worker health and attitude, community relations, competitive advantage and "doing the right thing" all contribute to the benefit side.

Respecting the tide of history

This point in industrial and human history is a convergence of ideas whose time has come.

- Health and safety of workers have received some formal guarantees under the law.

- ❍ Biosphere protection as defined by EPA is understood as an imperative.
- ❍ Human potential has become the highest valued natural resource promoting team building and problem solving.
- ❍ Information systems give more options for controlling waste and information on eliminating it.
- ❍ Total quality management shows what kind of waste is now unacceptable; cost cutting by eliminating waste becomes irresistible and everyone gets involved.
- ❍ Reengineering of corporations makes sleekness a value; a negative environmental history becomes a liability.

The Wheel of Foresight

These spokes turn the wheel of the world's industries toward a benign-by-design retooling with a high quotient of human creativity and an implied stewardship of the planet's resources, perhaps even a real movement toward restoration. This combination of historic *awakenings* converging in our lifetime might be defined as *foresight* for future generations.

This wheel will turn us toward a better world of both human and natural equality, sensitivity and stewardship. If we were all to put our shoulders to this wheel, the world could look very good indeed. Our environmental progress so far has been hard won, but humankind has, along with the rest of the life systems on the planet, paid a dear price for this awareness.

We now know the past. We know we can't afford to repeat it, nor do we want to. We can't go back to the waste-filled processes of this century or continue with the wholesale dissipation of our natural capital. We can only proceed to a future that promises to be centered around care and attention to the natural systems on which we depend. To do that, we must each ask ourselves:

- ❍ Do our actions maintain the fertility of the ecologic system?
- ❍ Do our actions complete the energy flow back to the earth?
- ❍ Does our organization build up our natural capital?
- ❍ Does it respect the limits of life on this planet?

A "yes" answer to these questions will change the way we do business in the future. Hang these questions on the boardroom wall and in every office where decisions are made affecting the environment.

Ultimately, choices in vision, mission, strategy and management come down to answering yes or no to these four questions. Long- and short-term economic goals must incorporate these considerations at every level if an organization is to behave responsibly as well as competitively. If future generations are to inherit the kind of planet we have enjoyed, maintaining and enhancing the biotic community must become as second nature as harvesting it has been. Good management requires it. In good conscience, we can do no less.

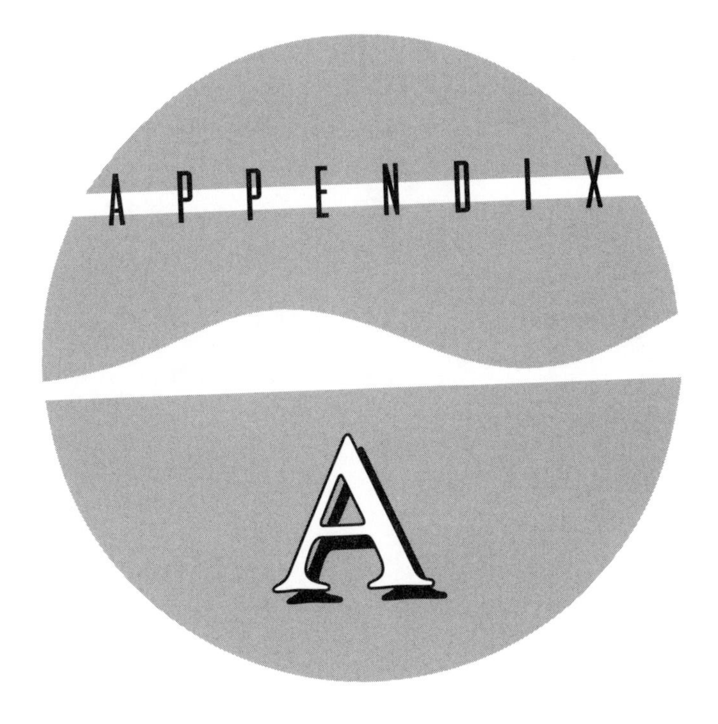
APPENDIX
A

American Airlines to Xerox

American Airlines
American Express
Anheuser-Busch
Bayer
Bell Atlantic
Chrysler Corporation
The Coca-Cola Company
Coors Brewing Company
Dow Chemical
DuPont
Eastman Kodak
Ford Motor Company
H.B. Fuller Company
General Electric
General Mills
General Motors
Gillette Company
Good Humor-Breyers
Hallmark
Hoechst Marion Roussel
S.C. Johnson Wax
Laidlaw
3M
Monsanto
JC Penney
Procter & Gamble
Sun Oil
Target
Walmart
Xerox

The following information was made available by each company through their annual environmental report.

American Airlines

Fort Worth, Texas

In 1993 American Airlines published its first Environmental Annual Report. Organized around regulatory lines, the report covers three environmental challenges: noise, recyclable refuse and harmful chemicals. Here are some of their successes.

Hazardous Waste

American's aircraft are unpainted and have been since the early 1930s. This factor reduces both weight and hazardous chemical handling. Where the aircraft's body is painted, the company uses the lowest in volatile organic compounds, uses water-based paints when possible and is developing acceptable methods for stripping paints and other coatings.

The company has switched to a water-based cleaning process for aircraft and engine parts formerly cleaned with trichlorethane, a chlorinated solvent that emits hazardous vapors. Consolidating use of degreasers combined with the water-based process has reduced the number of degreasers used to seven in 1994 compared with 40 in 1990.

Noise and Fuel Efficiency

Nearly 83 percent of American's fleet is Stage 3 aircraft compared to a national average of 57 percent, making it the lowest in noise generation and subsequently "most fuel-efficient fleet of all U.S. carriers, by far."

Underground Storage Tanks

American Airlines has invested more than $14 million to remove and upgrade underground storage tanks at 18 airports in anticipation of meeting EPA 1995 guidelines for bringing these facilities under compliance.

Recycling

An on-board recycling program begun in 1989 has recycled more than 1.2 million pounds of aluminum cans. The program has expanded to include newspapers, books and magazines left on board. To stimulate use of recycled material the airline has doubled its spending on products with recycled content since 1991.

At its headquarters in Fort Worth, Texas, the company operates a full-scale recycling program. The original investment of $166,000 for training materials, waste containers, can crushers, compactors and dumpsters has generated more than $115,000 in revenues and saved $38,000 in landfill disposal fees. The company estimates it has kept nearly 33,000 pounds of aluminum and 1,400 tons of paper out of landfills. This means it has saved 24,000 trees and almost one billion gallons of water. Ultimately, it has reduced office waste by 32 percent at its headquarters.

Community Outreach

American has formed a partnership with the Nature Conservancy. One project is the purchase of 43 acres of land near Nashville, Tennessee, funded by the flight attendants' can-recycling program. This area is home to the endangered Tennessee coneflower.

Awards

The company was recognized by the EPA in 1993 for the pollution-prevention benefits of not painting the aircraft, which may become an EPA option passed on to others in the aerospace industry.

American Express

New York City, New York

The company has undertaken an educational campaign for employees in the United States with an emphasis on recycling. It has an Environmental Protection Committee that recommends changes and policies.

Recycling

The American Express Tower in New York City saved about $250,000 by diverting recyclables from the waste stream. In 1993 the company recycled 4,337 tons of paper for a savings of $434,000.

Energy Conservation

An energy audit in 1992 has fostered adjustments at many facilities. It has joined Energy Star Buildings Pilot Programs and provided data to the EPA on the savings to be gained by using variable-speed drives for fan controls in an office building's heating, ventilating and air-conditioning systems. It is also a member of EPA's Green Lights Program, which promotes energy conservation in lighting.

Paper Reduction

Moving toward the paperless office, the company has introduced optical scanning equipment to reduce paper consumption and storage space and to increase work efficiency. Approximately 500,000 pages can be stored on one disk.

The use of e-mail is also being expanded and improved, which will further reduce paper use. An advanced report and management and distribution paper system in the Frankfurt, Germany, facility has reduced printed paper output by 35 percent annually since 1992. Improved document security and reduced work time and document-processing time are byproducts.

Community Outreach

American Express Foundation in the Czech Republic is supporting the building of a network of "greenways" that will connect historic towns and let tourists walk through undisturbed countryside near the city of Valtice.

It has published a book on tourism in conjunction with the Trust for Historic Preservation which disseminates information to communities on how to promote responsible tourism.

Employees have taken part in local campaigns, notably in Phoenix, where they helped to beautify, revegetate and build new trails in Papago Park, the Desert Botanical Gardens and the zoo. American Express has also participated in an annual Reef Sweep off the coast of Ft. Lauderdale, Florida, where divers cleaned up trash and debris, and in Toronto, where more than 100 employees took part in planting 700 trees and shrubs in October, 1993.

Employee Relations

The company circulated a questionnaire to employees on their views of environmental program needs within American Express.

Anheuser-Busch

St. Louis, Missouri

Anheuser-Busch dedicated itself to environmental concerns in 1970 when August A. Busch, Jr., pledged that the company would commit its resources to work toward solutions to environmental problems, particularly litter and solid waste. That early concern was expressed in a document called *A Pledge and A Promise*. The company still uses this as a theme in environmental efforts today.

Over the past several years the company has recycled one can for every can used to package beer. That amounts to 17 billion cans per year, more than 600 million pounds of aluminum. In 1992 the company's packaging contained 1.05 billion pounds of post-consumer materials. If this load were placed in tractor-trailer trucks, it would fill enough vehicles to form a convoy from Washington, D.C., to New York City, the company estimates.

Anheuser-Busch adopted a companywide waste reduction goal of 40 percent in 1990. In 1991 it reduced the waste stream by 11 percent. By 1993 it had reached 27 percent and was approaching 40 percent by the end of 1994. It also has a thriving wildlife and habitat protection program.

Recycling

A-B, one of the largest packagers in the world, purchases more than $2 billion worth of packaging materials containing recycled content. These purchases keep one billion pounds of materials out of the landfills annually. It has purchased more than 100,000 pounds of recycled letterhead and envelope stock and 5,000 pounds of business cards. It uses recycled computer paper with 10 to 20 percent post-consumer waste. The company uses more than 50 million sheets per year. Corporate office paper is recovered at an 80 percent level.

Diatomaceous earth used as a polish filter for the beer is recycled as an ingredient in cement. In 1993, four breweries diverted a total of 6,000 tons. In 1994 it expects to divert 10,000 tons of material by two-thirds of the participating breweries.

Beechwood chips needed in brewing used to be landfilled. In 1993, approximately 5,500 tons of chips were diverted by 11 breweries for use primarily in composting or mulch.

Plastic strapping holding the cases of beer on wood pallets had been tossed. The strapping is now chopped and returned to the manufacturer to be reformulated. In 1993 the company recovered 800 tons of this material.

Used gloves get a new life by being sent out for cleaning, sorting and repair. They are now recycled four or five more times around the plant before being discarded.

Benign by Design

In 1992 the company reduced the diameter of its can lids by 1/8". This reduces aluminum consumption by more than 21 million pounds per year. The energy saved from this reduction could provide enough electricity to power all the homes in St. Louis for five weeks. The electrical consumption of the company's can manufacturing subsidiary has dropped by 11 percent the last five years. The amount of water needed to wash the cans has dropped 45 percent in the last six years.

The newest brewery in Cartersville, Georgia, uses the latest environmental technology available. It treats the wastewater in the brewery's own bioenergy recovery system before releasing it, at greatly reduced strength, to Cartersville's municipal treatment plant. In the bioenergy recovery system, bacteria break down the organic content and create methane gas. The gas, used as a boiler fuel, reduces the need for other fuel by 15 percent. The same technology is on line at two other breweries and will be added to three others.

Suspended materials from the wastewater go by truck to a resource recovery farm where they fertilize hay crops. Spent grain is sold as cattle feed and has been since the turn of the century.

Water Conservation

Water conservation became a priority during the 1980s, particularly in California during the drought. Combined efforts at breweries have reduced by 17 percent the water used per barrel of beer, a savings represent-ing the needs of 28,000 households. A new system for washing the beechwood chips has saved more than 10 percent of the water used at some breweries.

A new, low-volume irrigation system at Cypress Gardens, California, has reduced water usage by 60 percent since 1980, although the number of plants and flowers has increased. At Sea World parks, filtration systems clean the water in animal pools so it is used continuously.

Energy

The energy needed to produce a barrel of beer has dropped 13 percent since 1984.

Awards

The company has received awards for waste minimization, community outreach and recycling.

Bayer Corp.

Pittsburgh, Pennsylvania

Bayer is a chemical manufacturing company that manufactures products ranging from crop-protection compounds to car parts to pharmaceuticals. Like other companies in this list, it has moved beyond compliance into voluntarily achieving higher standards. Bayer has created WRAM (Waste Reduction and Management), a program to eliminate or prevent waste at the source or to reuse or recycle to the maximum. WRAM teams meet during work time to recommend improvements. Since 1986, when EPCRA was passed, WRAM programs have reduced waste reportable under the Toxic Release Inventory. The company now works with EPA's 33/50 Program, a voluntary system to reduce emissions for 17 specified chemicals by 33 percent by 1992 and 50 percent by 1995. Bayer's goal in the chemical divisions was to reduce overall waste generation by 50 percent in 1993.

Through 1993 Bayer achieved a 44 percent reduction of all waste streams, including a 24 percent decrease in gaseous emissions, a 27 percent decrease in wastewater organics and a 45 percent reduction in solid waste generation. Other improvements include:

- A new plant in Baytown, Texas, will eliminate air emissions of toluene, used as a washing solvent, from polyol production. The plant will also reduce its solid waste by 95 percent.

- At its New Martinsville, West Virginia, plant, a major producer of hydrochloric acid for the steel industry, Bayer has installed a waste-gas thermal oxidizer to eliminate emissions with 99+ percent efficiency. This installation will virtually eliminate low concentrations of organics emitted in the atmosphere.

- A 55-acre lagoon, home to alligators, fish, birds and other animals, is maintained at the Bayer plant for Fibers, Organics and Rubber Division facility in Orange, Texas.

- In 1989, the first above-ground wastewater treatment facility in the United States was installed in the company's Baytown, Texas, plant. Called tower biology, the technology imitates nature's own water-purifying process by using enhanced oxygen absorption and naturally occurring bacteria to purify wastewater. Because of the above-ground tanks, the process virtually eliminates the potential for groundwater contamination. Emissions of aerosols and odorous fumes are avoided by the closed-system design.

Community Outreach

Safe handling of hazardous materials spills is a training program offered by Bayer in 1993 to more than 3,000 emergency-response personnel in communities around the country. Rail tank car and tank trailers filled with water give trainees a hands-on experience of a high-risk situation in a no-risk environment.

The company sponsored an annual Ohio River cleanup in 1991 for which Bayer donated time and equipment. Employees and community members removed 13,000 tons of garbage along 2,000 miles of shoreline of the Ohio River.

Bayer funds pre- and postdoctoral researchers at the University of Notre Dame's Center for Bioengineering and Pollution Control.

Bayer funds many small, local environmental projects from individual facility budgets.

Bell Atlantic Corporation

East Orange, New Jersey

Bell Atlantic recycles, partly because it is located in several states that have passed stringent recycling laws. The company's goals for recycling are ambitious.

- 50 percent reduction in trash-removal costs by 1994
- 80 percent reduction in volume of trash from all garages and work centers by 1995 (baseline 1992)
- 100 percent internal publications printed on recycled paper by 1994.

Recycling

The company has saved about $320,000 a year in trash-removal costs, gaining $30,000 in revenue from the recycling company. It has also created jobs for members of the Association of Retarded Citizens, whose members sort the company's paper for recycling. Plans are to expand this program. The company estimates it has saved $600,000 in hauling fees in 1992 and $690,000 in 1993.

Bell Atlantic published 34 million phone books in 1994 using 51,000 tons of paper containing a minimum of 24 percent postconsumer recycled fibers. This saves more than 10,000 tons of virgin paper and creates 10,000 tons of post-consumer waste paper, which fuels the recycling process.

Surplus or Obsolete Equipment

The company processes about 65,000 tons of surplus or obsolete telecommunication equipment. Some 30 percent of this is redeployed to other parts of the company to be reused. Marketable equipment is sold. The remaining 99.5 percent of the equipment is separated by metallic and nonmetallic content and converted to raw materials to manufacture new products. The goal is to convert the remaining half percent to recoverable use.

Awards

The American Forest & Paper Association recognized the company with a 1993 award.

Chrysler Corporation

Livonia, Michigan

Chrysler researches vehicles and the fuels they use as a unified system.

Alternative Fuel Vehicles

Chrysler produces natural gas-fueled minivans and full-size vans and will introduce compressed natural gas pick-up versions in 1995. A van equipped to operate on natural gas costs about $4,400 more than a comparably equipped gasoline vehicle. The driving range, about 150 miles, is limited to the space limitations of the on-board fuel storage canisters. A natural gas-powered fleet could be used where central refueling facilities are available, such as in trucks and vans for urban utility companies, small package carriers and airport parking lot transportation.

Flexible Fuel Vehicles

Chrysler is developing flexible fuel vehicle (FFV) technology to bridge the transition from having no fuel for alternative vehicles or having one kind of fuel but no vehicle that can use it. By offering a vehicle capable of operating on alcohol, gasoline or combination, customers can purchase such a vehicle in anticipation of fuel becoming available. The company has focused on developing M85, a fuel mixture of 85 percent methanol and 15 percent gasoline; however, the FFV could be modified to operate on ethanol-based 30S. These vehicles are required to meet the same tailpipe emission standards as gasoline-fueled vehicles.

The Dodge Intrepid will be delivered as an FFV in 1995 to retail customers in California. As refueling facilities become more widely available and consumer awareness increases, the market outlook could improve.

Electric Vehicles (EV)

Chrysler and Westinghouse Electric Corp. are conducting a joint development program to double the driving range between battery charges. With driving ranges now about 100 miles, much work needs to be done. The two companies are concentrating on an electric propulsion system with an AC electric motor and electronic controller that could make electric vehicles commercially viable by the year 2000. Their driving range would be as much as 200 miles.

The company offered its Dodge Caravan Electric and Plymouth Voyager Electric to fleet customers in 1994. A Chrysler/Nordvik charging system developed by Nordvik Technologies quickly charges any electric vehicle battery without overcharging and can be used in any weather condition.

Pollution Prevention

The company has emphasized "root-cause" pollution elimination. Here are some of their successes:

- Elimination of hexavalent chromium from all materials and processes
- Reformulation of paints and solvent cleaners to exclude the majority of listed toxic solvents
- Elimination of lead from all paints except electrocoat primer
- Reformulation of air conditioner refrigerants
- Phase out PCBs from Chrysler facilities by 1998

Recycling

Chrysler recycled eight million gallons of used oil from production processes. It has eliminated 55 percent of expendable packaging wastes from assembly plants by using durable returnable containers. It plans to eliminate 95 percent of packaging waste.

Thousands of tons of wood pallets and cardboard are recycled and 700,000 tons of scrap metal are salvaged each year.

The company has one of the largest paper-recycling programs in the United States, recycling more than 800 tons of paper per year.

Energy

The company has cut the energy used per vehicle built by 40 percent in the last 10 years through more energy efficient equipment and energy conservation programs.

Awards

Chrysler received the Stratospheric Ozone Protection Award from the EPA for its leadership in eliminating CFCs from vehicle air conditioning systems and facility processes.

All 1994 Chrysler vehicles use the CFC-free refrigerant HFC 134a in their air conditioning systems, including CFC-free polyurethane foam material in child seats for the Dodge and Plymouth minivans.

The Coca-Cola Company

Atlanta, Georgia

The company has a many-faceted program. Perhaps the most outstanding is its employee education program that has become a model for other companies.

Recycling

The company has developed a two-liter PET bottle made with 25 percent recycled plastic, which is used in international markets. The company explores new technologies for recycling PET into food-grade containers. Each syrup branch of the company diverts from landfills at least 80 percent of all solid waste generated. In the United States, recovered soft drink syrups are sold to recyclers who convert the sugars into usable fuel additives.

Recycling rates for soft-drink packaging are among the highest of any consumer products packaging. For example, in 1993, consumers recycled 63 percent of all aluminum cans, 35 percent of glass bottles and 41 percent of PET soft drink bottles.

The Atlanta headquarters diverts an estimated 750,000 pounds of material annually from the office waste stream, including office paper, computer paper, telephone books, laser printer cartridges, polystyrene cups and food cartons, aluminum cans, plastic cutlery, plastic bottles, newspapers, glass and corrugated cardboard. More than $100,000 has been earned from the sale of these materials to date, which has been donated to local charities.

Source Reduction

Coca-Cola has reduced the amount of raw materials used to produce its packaging. For example, glass bottles, aluminum cans and plastic bottles use 43 percent, 35 percent and 21 percent fewer raw materials, respectively, since their introductions.

Building Markets for Recycled Materials

The company uses recycled material in its primary packages, which exceeds 50 percent in aluminum cans and 30 percent in glass bottles. It has implemented a purchasing policy giving preference to products with recycled content and works with suppliers to identify and develop products containing recycled content.

Employee Education

Coca-Cola's comprehensive Environmental Development Program helps employees understand the issues so they may conduct business in accordance with the company's fundamental environmental principles and do their part to protect and enhance the quality of the environment. The program has become a model for other companies and was offered to customers and suppliers.

Community Outreach

The company works with Keep America Beautiful to improve waste-handling practices in communities across the United States.

Awards

The Coca-Cola Co. has received special recognition for its worldwide environmental management system with the following awards.

1994 *Earth Summit Award toward Global Sustainability*
United Nations Environmental Program, Center for Resource Management, and the Earth Council

1993 *The Stratospheric Ozone Protection Award*
United States Environmental Protection Agency

1993 *Award for Outstanding Corporate Environmental Achievement*
National Environmental Development Association

1992 *Award for Technology Achievement in Environment Category*
Discover magazine

Coors Brewing Company

Golden, Colorado

The company published its third environmental progress report in 1994 in which it describes early innovations: Coors introduced modern waste-water treatment in Colorado in the 1950s and the two-piece aluminum can and can recycling in 1959. In 1988 it responded to increased environmental interest and legislation by committing to corporate environmental responsibility, a position that required greater compliance to regulations and voluntary ways to pollute less.

Emissions

Chemical emissions reportable under the EPCRA and Community Right-To-Know Act declined by 15.5 percent in Golden in 1993, primarily through process changes at the can-manufacturing plant (the result of employee-initiated pollution prevention programs) a trend that has resulted in a 57 percent reduction since 1989.

- In 1990 the company emitted 14 chemicals requiring tracking for a total of 533,107 pounds. In 1993 the company had reduced that to five chemicals at 261,154 pounds.
- The brewery in Golden reduced ammonia releases by 86 percent in 1993.
- The company conducted a voluntary assessment of volatile organic compound emissions and found government estimates of these figures to be low. The study is changing the understanding and procedures to manage these emissions for brewers all over the country.

Recycling

- The company has launched SCRAP—Save, Conserve, Recycle and Profit, and program that assists employee teams in identifying, evaluating and implementing projects that provide both environmental and economic benefits and recognizes these efforts with citations and awards.

Results have been achieved in the downweighting of aluminum, glass and paper packaging, purchasing of office supplies made of recycled materials, recycling of office paper, selecting of recycled paperboard and soy-based inks for point-of-purchase materials.

- Coors is one of the companies to accept the CONEG Challenge, a call by the Coalition of Northeastern Governors to reduce product packaging waste
- The recycled content of aluminum cans reached 67 percent in 1993
- In 1987 the company sent 76,360 cubic yards of nonhazardous waste to landfills; by 1993 it had reduced that to 30,747 cubic yards
- Its postconsumer recycling of glass has tripled since 1989 to 51,429 tons; the annual average for recycled content has climbed to 33 percent with 27 percent postconsumer waste
- Coors has built a $4.5 million recycled-glass-processing facility in Wheat Ridge, Colorado, which has doubled the capacity to 100,000 tons per year.

Corporate Culture

Peter Coors signed the company's Environmental Principles in 1992 and expanded them in 1993 to formalize existing efforts, some of which include:

Recognize employee accountability for controlling environmental impact through the Performance Appraisal System

Recognize the "priority of environmental factors in achieving economic goals"

The company created a position of Chief Environmental, Health and Safety Officer.

Community Outreach

Coors surveyed its stakeholders in 1992 to see how these groups perceive corporate environmental responsibility in general and Coors's response in particular. The survey was designed to gather input and direction. The survey showed that stakeholders felt Coors could work harder at compliance and create voluntary programs that further reduce the company's effect on the environment. Respondents agreed the corporations have an important role in protecting the environment and that that should be a top priority. The company is continuing the dialogue and acting on the recommendations.

Pure Water 2000, launched in 1990, identifies and supports local, grass-roots water protection efforts. The company has provided more than $2 million in support of more than 600 local water protection and education projects. Working through its distributors, the company funded more than $535,000 in cleanup projects in 1993.

Dow Chemical

Midland, Michigan

Dow created an employee involvement program in 1986 called WRAP (Waste Reduction Always Pays). Since that time both individual employees and employee teams have worked to eliminate unnecessary waste and have discovered that waste reduction pays off.

Recycling

Improvements have ranged from the extremely technical to the simple and obvious. Results for one WRAP project alone are an 80 percent reduction in its latex plant's waste to the landfill, a 50 percent reduction to the incinerator and a 78 percent reduction in air emissions. These changes saved $700,000 a year in environmental costs and improved yield. The company has now instituted the same procedures in a new plant in Thailand. Overall, the WRAP program has reduced Dow's wastes and emissions by an estimated 120 million pounds per year.

More than 90 percent of the 7,400 employees at the Dow headquarters and plant in Midland, Michigan, participate in a voluntary paper-recycling program. Beginning Earth Day, 1990, the program has recycled more than five million pounds of office paper for a net savings of $650,000 per year. In 1991 the WRAP program started recycling polystyrene food service and packaging materials and now has recovered and recycled more than 3,000 cubic yards of polystyrene.

Emissions

Volatile organic compounds are removed from vent streams at the rate of 99.9 percent by a solvent vapor recovery unit that exceeds rates of 95 to 98 percent for conventional systems. Twenty such units are now in operation with 20 more under development throughout the world.

Dow has closed the loop in recycling and reusing waste water in its plant in Fort Saskatchewan, Alberta, Canada. Impurities are removed by a large evaporator that boils the waste water. Nontoxic mineral solids go to an underground disposal cavern; the water goes back into the plant for reuse. New plants in Thailand and Indonesia incorporate developments such as ventless polyol reactors to eliminate air emissions and zero process water discharge for polyol and latex plants. Polyol, a basic ingredient of polyurethane plastic, is used in the production of auto parts, shoe soles, molding, insulation and other products.

CFC Alternative

In October 1993 Dow Plastics received patents for its 100 percent carbon dioxide blowing agent technology, replacing CFCs and HCFCs in the production of thin polystyrene foam sheets for food packaging. This process is used make bowls, cups, egg cartons, meat trays and cafeteria trays with no harm to the ozone layer. Dow has licensed the technology to six companies. The full conversion of these licensees will eliminate the use of more than three million pounds of HCFCs per year.

Styrofoam

CFCs were phased out of styrofoam brand insulation in June 1990. The company now uses HCFCs instead, which it says are 95 percent less ozone-depleting; it is searching for alternatives to them. In corporate partnerships, Dow uses plastic lumber to make floating docks. The lumber contains mixed recycled plastics and some scraps of styrofoam.

Community Outreach

Dow Plastics, in partnership with the White Sox, the Solo Cup Co. and others, initiated a polystyrene cup recycling effort at Chicago's Comiskey Park. White Sox fans recycled more than two million polystyrene cups in 1993. This is the first full-season polystyrene recycling program in a U.S. major league stadium.

Dow also works with the National Park Service since 1989 to recycle nearly 3,000 tons of recyclables at seven national parks, including the Grand Canyon and Yosemite. The company has assisted in creating a recycling infrastructure for each park and education materials for visitors. Dow has now coauthored with the Park Service a recycling resource guide summarizing the lessons learned from the seven participating parks.

DuPont

Wilmington, Delaware

DuPont has published its second environmental report and declares it is committed to "corporate environmentalism" around the world. Some of the successes of the past five to seven years include:

- A phase-out plan for CFCs, targeted for 1994 but slowed because of a request by the government for more time to develop alternatives. Production levels continue to decline to less than half of 1986 levels
- Reduction of air toxics by 65 percent and airborne carcinogens by 70 percent since 1987
- A 43 percent reduction of 17 large-volume toxic chemicals from 1988–1992
- A 20 percent drop in energy use per pound of product
- A 3.7 percent decrease of carbon dioxide emissions from manufacturing
- Four double-hulled tankers in operation, a feature now true of more than half of their fleet
- More than 120 double-walled underground storage tanks installed at Conoco service stations
- 30 percent of packaging returned for reuse and recycling
- Plans to reduce waste by 25 percent in 1994 and 50 percent in 2000
- A new process for dyeing nylon textiles reduces dye and chemical requirements by 25 percent, water and steam consumption by 50 percent and dye discharges tenfold
- Awarded employees with Environmental Excellence Awards for five years

- A carpet reclamation program diverts more than two million pounds of used carpet from landfills to date.
- Sulfonylurea herbicides have been developed that use ounces per acre (rather than pounds) and control weeds effectively.
- 100 million pounds of waste was recycled in 1992 with energy recovery value of 80 million pounds.

Employee Awards

DuPont encourages its employees to create solutions in environmental areas and it recognizes performance with an awards program. Twelve recipients are selected annually for awards from usually more than 400 nominees. The award winners designate environmental organizations to receive grants that total $60,000 each year. More than $300,000 has been awarded since the program began.

Corporate Culture

Terms such as "product stewardship," "life-cycle analysis" and "zero waste" indicate the company's commitment to progress. DuPont has committed to zero-waste generation at its source, a full phase out of CFCs, a 50 percent reduction of packaging waste, a 90 percent reduction of carcinogenic air emissions and a 35 percent reduction in the manufacture of hazardous waste by the year 2000. The chairman defined corporate environmentalism in 1989 as "an attitude and performance commitment that place corporate environmental stewardship fully in line with public desires and expectations."

New Products

DuPont has several new products in the works, including a process to expand polyester recycling capabilities. It plans to convert postconsumer and postindustrial materials once thrown away to produce virgin quality raw materials without the oil-derivative feedstock necessary for new polyester production.

Investment

The company will invest $12 million to convert an existing dimethyl terephthalate (DMT) production facility at Cape Fear, North Carolina, to a methanolysis facility, to use waste polyester as a feedstock in place of araxylene. The facility will be operational in 1995.

Feedstock

DuPont plans to retrieve the unused polyester film from customers through a film recycling program. The facility will initially produce 100 million pounds of DMT and 30 million pounds of ethylene glycol annually. Responding to customer interest in buying apparel containing first-quality recycled polyester, the new methanolysis facility will help it respond to that demand.

Product

The product will be first-grade polyester, including Mylar polyester film, Dacron fibers, Cronar polyester base film and Crystar polyester resins.

Eastman Kodak Company

Rochester, New York

In its fifth annual environmental report outlining results for 1993 the company mentions a wide array of successful programs to boast about. Nine Kodak Guiding Principles speak to the corporate commitment to health, safety and environmental responsibility. They are worthy of being quoted here:

Principle 1

To extend knowledge by conducting or supporting research on the health, safety, and environmental effects of our products, processes, and waste materials.

Principle 2

To operate plants and facilities in a manner that protects the environment and the health and safety of our employees and the public, and is efficient in the use of natural resources and energy.

Principle 3

To make health, safety, and environmental considerations a priority in our planning for all existing and new products and processes.

Principle 4

To develop, produce, and market products and materials that can be manufactured, transported, used, and disposed of safely and in a way that poses no undue environmental impact, and to provide services in a safe and environmentally sensitive manner.

Principle 5

To counsel customers on the safe use, transportation, storage, and disposal of our products, and for those services we provide, to provide them safely.

Principle 6

To participate with governments and others in creating responsible laws, regulations, and standards to safeguard the community, workplace, and environment and in applying environmentally sound management practices and technologies.

Principle 7

To measure our environmental performance on a regular basis and provide—to officials, employees, customers, shareholders, and the public—appropriate and timely information on health, safety, or environmental hazards, initiatives and recommended protective and preventive measures.

Principle 8

To recognize and respond to community concerns about our operations and to work with others to resolve problems created by handling and disposal of hazardous substances.

Principle 9

To encourage employees to apply off the job the same principles of health, safety, and the environment that are applied at work.

Kodak has reduced emission by about 30 percent since 1988. As a participant in the EPA 33/50 program, Kodak achieved a 54 percent reduction of targeted chemicals from 1989 to 1992, exceeding both the program's and the company's goals. Kodak CFC emissions were reduced 57 percent worldwide, exceeding their goal of a 50 percent reduction by 1993. More than 10.5 million cameras (3 million pounds of plastic) were diverted from landfills in 1993 through their single-use camera collection and recycling program. "Sustainable development means growing in harmony with the earth," says Hays Bell, Vice President of Corporate Health, Safety, and Environment. The company seems on its way to doing just that and has created hundreds of changes, both large and small, toward "growing in harmony."

Ford Motor Company

Dearborn, Michigan

Ford has undertaken several environmentally sound steps.

Alternative Fuel Vehicles

Ford is working with the U.S. Department of Energy and seven other advanced technology companies to develop a hybrid electric vehicle. The $133 million research effort is 50 percent funded by the Department of Energy and supports the goals of the SuperCar initiative, the national Partnership for a New Generation of Vehicles. The hybrid's engine refuels rapidly and offers a long driving range. Three types of hybrid electric vehicles will be constructed: range-extended electric, series and parallel.

Ford also will be working on a separate, $13.8 million program to develop proton-exchange membrane cells fueled directly by hydrogen to achieve a powertrain with zero emissions, although all hybrids are not necessarily zero emissions. This fuel cell development will complement the hybrid electric vehicle program that began in 1993.

Ford Ecostar

Ford is building 103 electric vans as part of an international program to put the company's advanced electric vehicle technology to the test. This is the nation's largest test fleet of electric vehicles. These vehicles will be leased to utility and delivery companies around the country for 30 months for testing. The van has an advanced sodium sulfur battery with a 100-mile range, advanced computer-based electronics and other features to make it very efficient. Load carrying capacity is 900–1000 pounds.

Tire Recycling

Tires are recycled into new parts, including brake-pedal pads and floor mats for new Ford vehicles in 1995. One tire can create 250 pads made with 50 percent recycled tire content. Ford is the largest automotive user of tire scrap; it purchases more than 18 million pounds each year. Using old tires is a challenge, because once the material is heat-molded and shaped, the chemical makeup of the rubber changes, making reuse very difficult.

Recycled Parts

Ford is looking for ways to increase the use of used vehicle parts. Although 75 percent of a vehicle is recycled already, Ford is working to recycle the remaining 25 percent. Some things the company does include:

- Plastic bumpers have been recycled into tail lamps on Taurus and Sable wagons. Each tail lamp contains at least 50 percent recycled material.
- Grille assemblies, luggage racks and door padding are made from 50 million recycled plastic soda bottles per year. Ford is the nation's largest automotive user of this material.
- Old plastic battery housing has been turned into new splash shields.
- Plastic wrap from assembly plants is now made into protective seat covers, which saves more than 400,000 pounds of plastic from landfills.
- Ford is recycling plastic water bottles—like those used in office water coolers—to make new headlamp housings for the 1995 Explorer and Ranger.

Community Outreach

Ford has supported five national parks since 1990 and is considering others. Its Wildlife at Work project establishes wildlife enhancement sites on corporate land at an assembly plant in Cuautitlan, Mexico. The project demonstrates sustainable development and compatibility of industry and the environment. This project is the first in Mexico, although 300 wildlife sites have been established in the United States, Canada, Spain and Australia. Ford collected 250,000 Christmas cards that were then donated to St. Jude's Ranch in Boulder City, Nevada, where they are remounted and sold as new cards.

Awards

The Society of Plastic Engineers recognized Ford's innovative use of plastics in the environmental category in 1993.

H. B. Fuller Company

Saint Paul, Minnesota

H. B. Fuller emphasizes prevention rather than remediation because "it's the right thing to do." The company manufactures and markets adhesives, sealants and coatings, paints and other chemical products. It also produces powder coatings to metal finishing industries, commercial and industrial paints. A small company compared to others in this book, H. B. Fuller employs 6,000 workers in 32 countries. In 1993 the company was the first specialty chemical company to endorse the CERES Principles. These 10 concepts of environmentally sensitive corporate behavior, originally named the Valdez Principles, were an outgrowth of that tragic incident in 1989. CERES stands for Coalition for Environmentally Responsible Economies and is a nonprofit group in Washington, D.C., that promotes these principles to companies around the country.

Pollution Prevention

Motivated by federal Toxic Release Inventory requirements, H. B. Fuller chose to use the Act to measure pollution prevention efforts. Fuller participates in EPA's 33/55 program; its goal was to reduce the targeted chemicals by 70 percent rather than the suggested 50 percent by 1995, but it achieved a 72 percent reduction by 1993—two years ahead of schedule. The company removed trichlorethane, an ozone-depleting chemical, from its water-based products manufactured by North American Adhesives, Sealants and Coatings Division in 1992. It phased out use of the chemical as a raw material and cleaning solvent in all Fuller operations in 1994.

Waste Water Prevention

Waste water from production processes is recycled and reused wherever possible, helping to reduce the consumption of fresh water as well as the volume of waste water. Internal standards of the company require production and warehouse facilities with direct discharges to surface waters to have waste water treatment systems.

Waste Disposal

Using 1989 as a baseline, the company set a goal to reduce waste by 50 percent by the year 2000. To reduce solid waste and conserve raw materials, it reworks scrap material into the manufacturing process and reclaims and reuses solvents in the production process.

Fuller has increased its use of returnable tote bins and crates from 52 percent in 1991 to 56 percent of all tote sales in 1992. A flexible bulk container system that holds as much as 1,600 pounds of hot melt adhesive will help to eliminate excess packaging, strapping and stretch wrap.

Risk Management

Fuller's proactive approach is embodied in its use of Cameo (Computer-Aided Management of Emergency Operations). This database accounts for each of the company's facilities. Key contact names, bulk storage and equipment information and a diagram of each building and its general contents are all recorded in the database.

The program also models the accidental release of volatile chemicals. Each facility has its own spill simulation, projecting on a site map the direction, size and rate of movement of the plume that might result from a worst case scenario. This information is used in the company's emergency response plans. The program has been in place since 1987.

General Electric

Fairfield, Connecticut

General Electric is one of the oldest companies listed here—more than 100 years old with more than 200,000 employees—a mature elder of the Industrial Era. The company says it now wants to move from reaction to prevention.

Emissions

In the six years from 1988 to 1993, the company reduced its releases of 17 targeted chemicals by more than 73 percent in a voluntary emissions reduction program. It exceeded the target goal of 50 percent three years ahead of the EPA scheduled date. GE achieved a 64 percent reduction on EPCRA 313 emissions and reduced its release of freon 113 and methyl chloroform, two ozone-depleting chemicals commonly used by GE, by 96 percent and 83 percent respectively.

GE Plastics completed a voluntary program in 1992 in which it cut emissions by 75 percent from 1987. The chemical industry average is 36 percent for reducing emissions during that time period. This reduction means GE Plastics has prevented almost 60 million pounds of material from being emitted into the environment.

GE Aircraft Engines has used manufacturing modifications and materials substitutions for items such as cleaners and coolants. These changes have created a reduction in emissions of 80 percent—from 2.9 million pounds to 0.6 million pounds. It is working toward eliminating the use of all ozone-depleting chemicals from its operations.

Training

The company has required plant managers worldwide to attend a two-day environmental awareness training that highlights their environmental responsibilities. A company-wide prevention program called Power (Pollution, Waste and Emissions Reduction) has promoted pollution prevention since 1990.

Product Development

GE Appliances has invested more than $25 million to phase out CFCs in foam insulation and as coolants in its products. It has introduced refrigerators that are CFC-free before government regulations would require it.

GE Industrial & Power Systems produces power generation equipment considered to be the world's most efficient. A combination of steam and gas turbine technology in "combined cycle" reaches fuel efficiencies above traditional coal-fired plants. Other innovations include technology that converts coal to a clean-burning natural gas for electricity generation. This business is also a world leader in pollution control devices for stack emissions.

GE Lighting has developed nearly 200 types of energy-saving light bulbs in the last five years. Five of the products have received Green Seal certification for energy efficiency.

GE Motors has introduced electronic, programmable motors achieving a 20 percent efficiency increase over similar-sized conventional motors. These motors can be found in small appliances and air conditioners and in large industrial processes.

GE Plastics manufactures highly engineered plastics that can be used with corporate partners.

- Ford users GE plastics in its bumpers and then recycles them in other parts.

- One partnership connects McDonald's, DEC and Nailite, a manufacturer of building materials, to recycle computer housing made of GE plastics into roof tiles for McDonald's restaurants.

GE makes thermoplastic for recyclable milk containers for school lunch programs. After the bottles are used more than 100 times, they are returned to the manufacturer to be recycled into other products. The school saves the waste-hauling fees.

Awards

Recognition has come from Green Seal for energy efficiency, the first Fortune 10 company to receive this approval for its products. The EPA recognized GE Lighting in its Green Lights Program and gave the company its Certificate of Distinction Award. GE has received several state awards for pollution prevention and was honored by the U.S. EPA for its emission reductions under the 33/50 program.

General Mills Inc.

Minneapolis, Minnesota

General Mills so far has focused on the solid waste issue.

Recycling

General Mills has minimized the amount of material used in packaging, maximized the use of recycled packaging and packaging that can be recycled, used packaging materials that do not release harmful substances and used display codes and symbols to direct and inform the consumer.

Recycled material constitutes 98 percent of the cartons produced. All the paper in the dry-food cartons is recycled. At least 35 percent is from postconsumer waste.

The company's corrugated shipping boxes contain 35 percent recycled paper. These are often reused many times between plants before they are recycled. In 1991, 56 million pounds of corrugated boxes were recycled by General Mills plants and 20,000 pounds of plastic scrap was made available for recycling into other products.

Responsible Packaging

Reducing the weight of Hamburger Helper cartons reduced packaging by two million pounds of material per year.

A new tray-style shipping container for cereals has reduced packaging by more than two million pounds.

General Mills has reduced the weight of the Pop Secret popcorn bag and reduced packaging by more than one million pounds per year.

Gorton's fish packages have been repackaged in a lightweight film bag, reducing waste by 1.5 million pounds.

The company has reduced the thickness of the plastic liner in cereal boxes by 12 percent, reducing waste by 500,00 pounds per year.

General Motors

Warren, Michigan

Minimizing the environmental impact of GM products and operations is an integral component of the General Motors team vision. While GM products and manufacturing operations are subject to environmental requirements and expectations, GM views caring for the environment as more than a responsibility: it is a critical factor in GM's success.

Product Stewardship

A life-cycle perspective is fundamental to GM's commitment to product stewardship. Automobile life-cycle issues are complex, with environmental impacts occurring at raw material acquisition, manufacturing, use and disposal stages of product life. GM's Design and Manufacture for the Environment Committee serves as the coordinating body to apply the GM environmental principles to product design and related manufacturing processes.

GM has pursued alternative fuel vehicle technology for more than 25 years. The alternatives that have been considered include methanol, ethanol, natural gas, liquefied petroleum gases, electricity, hydrogen and coal. Each of these fuels has advantages and shortcomings compared to gasoline-fuel vehicles.

In 1993, GM began production of 50 limited production units of its electric-powered Impact, and announced it will team up with electric utilities to invite customers to drive them for two- and four-week periods, beginning in late spring 1994. This nationwide, two-year program, called the GM PrEView Drive, is designed to show individual motorists what electric vehicles can deliver, and for GM to learn from customers what they will require in a viable electric vehicle. In addition, GM's components divisions are actively developing electric vehicle components.

Pollution Prevention

The company launched WE CARE (Waste Elimination and Cost Awareness Rewards Everyone). The program asks every employee to "modify existing methods, procedures and processes and to incorporate waste prevention into all new endeavors." This program is a joint effort with the United Auto Workers as a continuous improvement initiative.

In 1992, GM facilities implemented projects that prevented, eliminated or reduced emissions and wastes going into the environment by approximately 318,000 tons. Many efforts are also under way at GM facilities to reduce dependence on landfills. Packaging reductions are achieved by working with assembly plant parts suppliers to reduce the volume of incoming packaging and to make packaging materials more recyclable. Twenty-five plants are involved, and they have reduced packaging from 82 pounds per vehicle to 15 pounds. Some plants generate as little as 1 pound per vehicle.

Environmental Resource Guide for GM Dealers

General Motors has provided an environmental resource guide, *Living with the Environment: A Resource Guide for GM Dealers,* to its 8,500 dealers. The guide contains practical information for improving environmental performance; the manual will be updated as information becomes available.

Community Outreach and Education

Since 1989, GM staffs and plants worked with the University of Michigan and local public schools to implement the Global Rivers Environmental Education Network (GREEN). Through Green, teachers all over the world are taking students to their local river and teaching them to monitor water quality, analyze watershed usage, identify the socioeconomic determinants of river degradation and present their findings and recommendations to local officials.

In 1993, GM and GREEN formed a partnership with the EPA and the U.S. Department of Energy to develop regional Green programs. The pilot project of this partnership is the Harpeth River Environmental and Education Project in the Williamson County School District, Tennessee. GM's goal is to initiate two new Green programs per year.

Changing Fuel in Power Plants

The GM Inland Fisher guide plant in Anderson, Indiana, used a coal-powered plant to power its operation. The coal ash-produced as a by-product became increasingly costly to handle and dispose. Not only did tons of it go to the landfill, but the plant paid surcharges for water treatment to clean the particulates from the waste water generated by the plant.

The plant boiler was changed to a natural-gas burning process in 1992, which eliminated the ash problem and the need to upgrade waste-water treatment facilities or add more controls to boiler stacks to control emissions. Particulate emissions fell from about 1,500 tons per year to zero. Sulfur dioxide emissions fell from about 1,200 tons to nearly zero and nitrogen oxides from 800 tons to about one. Moreover, the gas-fired boiler can be turned off on weekends, saving $9,000–$18,000 per weekend.

The Gillette Company

Boston, Massachusetts

Gillette manufactures and markets personal-care and -use products, including blades and razors, stationery products, Braun electric shavers and small appliances and Oral-B oral-care products. The company has more than 50 sites in more than 20 countries. In 1990 it created three internal task forces on packaging and plastics reduction, chemical solvent use reduction and emissions reduction. The company's Pollution Prevention Program represents a worldwide effort to reduce emissions to air, surface waters and municipal treatment facilities, as well as hazardous waste shipments to off-site locations for treatment or disposal.

Reducing Emissions

The Pollution Prevention Program aimed to reduce hazardous emissions by 50 percent by 1997, but the company achieved 50 percent reduction by 1994. Emission reductions are reported at nearly four million pounds or 59 percent of the 1987 base-year emissions rates. Gillette has installed aqueous wash systems to replace industrial cleaning chemicals such as TCE (trichloroethylene), TCA (1,1,1 trichloroethane), CFCs and PCE (perchloroethylene).

Hazardous Waste

Gillette wants to reduce the amount of hazardous waste shipped by its plants to off-site disposal facilities. In 1986–1993, Gillette reduced by 45 percent the volume of hazardous waste disposed off-site. During the same time the company's U.S. business grew more than 60 percent.

Chemical Solvents

Within the stationery products group, solvent content as a percentage of product formulation was reduced by 30 percent from 1989 to 1993. Company-wide, the number of listed solvents used in production quantities greater than 1,000 pounds was reduced by 35 percent.

The company plans that usage of all ozone-depleting chemicals in its products or processes will cease by year-end 1995.

Toxic Materials

When consumers dispose of old Braun rechargeable shavers, the battery cells enter the solid waste with the rest of the product. The company has developed an environmentally friendly battery called Green Cell, which is incorporated in electric shavers. Instead of cadmium, lead and mercury, the battery's energy source is hydrogen, which is stored in a special alloy in the negative electrode.

Plastics and Packaging

In 1990–1993, the North Atlantic Group, including North America and Europe, eliminated more than 10 million pounds of packaging by eliminating tuck boxes on personal care products, redesigning the Sensor shaving system secondary packaging to increase efficiency, and redesigning the entire White Rain hair care line. The North Atlantic Group reduced its annual plastic consumption by 1.7 million pounds through the redesign of the White Rain line.

Recycling

At the Prudential Tower headquarters, Gillette increased its recycling of waste paper by 14 percent in 1993. The company recycled 152 tons of white paper, the equivalent of 2,584 trees.

Water and Energy Conservation

Since 1972, the company has reduced by 96 percent the amount of water it takes to make a razor blade. This annual South Boston water savings of 650 million gallons equals almost a three-day demand on the entire Massachusetts Water Resource authority system serving the Boston area.

The company has been a member of the EPA's Green Lights Program since 1991 and has saved more than 8.8 million kwh annually on lighting alone at its domestic Gillette facilities since that time. In 1993, energy conservation projects in its U.S. facilities saved 13,500,000 kwh per year, equal to the amount of energy used in 27,000 homes in one month.

Awards

In 1990 Gillette received a Governor's Environmental Achievement Award from the Commonwealth of Massachusetts. In 1993 the company's World-wide Conservation Program won a National Environmental Achievement Award from Renew America, a national environmental organization. In 1994, Gillette was honored for meeting the goals of the Massachusetts Packaging Challenge, which calls upon companies to increase the recycled content of packaging materials.

Good Humor–Breyers

Framingham, Massachusetts

The Good Humor–Breyers manufacturing facility in Framingham, Massachusetts, built in 1963, produces approximately 20 million gallons of ice cream each year under the Breyers, Sealtest and Light n' Lively labels. By the mid-1980s, the facility was not competitive with other manufacturing facilities in its operating group, partly because it consumed 85 percent of a production day's electricity on a nonproduction day.

Working with the Boston Edison, Co., the Good Humor–Breyers facility underwent an energy audit conducted by the Energy Advisor Service, a program operated by the Commonwealth of Massachusetts Energy Office. As a consequence of measures taken from audit recommendations, the Good Humor–Breyers facility stayed competitive, Boston Edison kept a big customer and the program has become famous.

The Audit

The energy conservation opportunities identified for the Framingham facility were:

1. Replace a 1,700-ton freon refrigeration system with a 2,000-ton, three-stage ammonia refrigeration system, CFC-free, that has an automated control system.
2. Replace the air-handling equipment in the ice cream hardener and storage freezer. The new air-handling system would circulate twice the volume of air with half the horsepower and use the hot ammonia gas for defrosting rather than electricity.
3. Install new light systems (T-8 metal halide) throughout the facility except for the production area, which was already energy efficient.
4. Install 40 energy-efficient motors.
5. Install a heat-recovery system designed to save approximately 10,000 gallons of fuel oil and 100,000 gallons of water and waste water annually.

The first four recommendations were estimated to save six million kwh of electricity annually, or 33 percent or the 1988 electric consumption.

The Process

The Energy Advisor Service presented the recommendations in February of 1990. The Framingham facility accepted the audit report and, by deciding to implement all the recommendations, became the first Boston Edison customer to participate in the Energy Efficiency Partnership with the utility company. The next step was to persuade the Kraft General Foods, which owned Good Humor–Breyers at the time, to fund the $3.6 million project plus $250,000 to do the last recommendation. Kraft agreed to fund the project because the Boston Edison's incentive was based on verified energy savings. Funding was secured in six weeks. An aggressive schedule was developed to design and construct the project in 12 months. The completed project varies from the audit report because operating personnel modified the projects in the design and construction phase.

Results of Retrofit

12,545,528 kwh of electricity saved, or 112 percent of the targeted savings

A two-year verification period documented that refrigeration energy savings for the two-year period was eight million kwh compared to the estimated 7.3 million kwh. The air-handling system, motors and lighting savings were on target: 3,985,325 kwh savings for air handling and 187,000 kwh for motors. Lighting saved 284,477 kwh, a 30 percent reduction in lighting consumption. This verified savings qualified for incentive payments of $3.3 million from Boston Edison, resulting in a one-year payback for Kraft General Foods.

Benefits

Electricity, water and fuel are now viewed as ingredients of the ice cream manufacturing process and are analyzed daily. Since 1988 the kwhs per gallon of ice cream have decreased by 25 percent and electric cost per gallon by approximately two cents per gallon. During a period of high production in the spring of 1994, electric cost per gallon dipped below five cents a gallon.

Permitting employees to influence design and construction paid off when workers increased line efficiency by more than 10 percent since the first quarter of the verification period. The combined results of the Energy Efficiency Partnership and the increase in line efficiencies places Framingham second in the Division.

Update on the Energy Efficiency Partnerships

The audit, implementation and verification are the three major components of the retrofit program offered by Boston Edison's Energy Efficiency Partnership.

The audit identifies and analyzes electrical use within a facility. A comprehensive list of energy conservation opportunities have been screened by Boston Edison to determine which are cost effective and eligible for an incentive. If customers agree to undertake all of the eligible opportunities, Boston Edison assumes 100 percent of the audit's cost. If customers select some of the eligible recommendations, they must pay 50 percent of the audit cost associated with those not selected.

Customers assume the leadership of the project throughout the implementation phase, which consists of two components: design and construction. Boston Edison provides professional support throughout the process. It may turn out that when bids are received, implementing the recommendations will be cost effective; if not, Boston Edison will pay the design costs. The construction phase consists of the construction of operation and maintenance procedures and commissioning.

The verification phase is the critical phase because it determines the incentive the customer receives. The original verification phase was two years and the maximum incentive was the cost of implementing the recommendations minus the first-year energy savings. In 1992 the verification period was reduced to one year and the maximum incentive was reduced to the cost of implementing the recommendations minus a year and a half of the energy savings.

The retrofit program is fully committed through 1995, but its future is uncertain after that.

Hallmark Cards, Inc.

Kansas City, Missouri

In 1990 Hallmark formalized its environmental efforts, which included reducing waste, increasing use of recycled materials, reducing emissions and conserving energy.

Environmental Goals for 1995

Using the base line year of 1990, the company set these goals:

- Recycled paper use in products and offices will increase to 45 percent from 8.29 percent.
- Solid-waste disposables will decrease by 70 percent.
- Energy consumption will decrease by 15 percent.
- Volatile organic compound emissions from printing operations will decrease by 80 percent.
- Hazardous waste disposal will decrease by 50 percent.
- Packaging reduction projects should reduce materials by 5,343,400 pounds.

Hazardous Waste Disposal

Hallmark continues to explore the use of less hazardous materials, including water-based paints, printing and silk screen ink, cleaning materials, adhesives and aqueous developers for printing plates. These changes have reduced the quantity of hazardous waste that has been disposed over the past several years.

Reducing Waste

Hallmark has found that making small adjustments can result in big savings. One small merchandising redesign saved nearly 100 tons of paper a year and about $750,000. Formerly, after one die was used to cut a press sheet full of many different cards, the die was discarded after one press run. A new system creates a die for each card that can be reused grouped in different configurations. The new method saves the equivalent of $590,000 in materials and labor annually.

Emissions

Converting from solvent-based inks to water-based inks is an effective way to reduce volatile organic compounds. Solvent-based inks release small amounts of solvents into the air and produce waste that must go to a hazardous-waste treatment facility. About 85 percent of Hallmark's gravure and screenprinted products are now produced using water-based ink in such items as gift wrap, party plates, cups, stickers and banners.

Many hazardous solvents in cleaning products have been replaced with oils from citrus fruits. Employees find that these can be more effective than the chemical products previously used. These new formulas also reduce air emissions, hazardous waste and fire hazards.

Energy

Lighting has been modified and glare reduced under the recommendations, resulting in a 225 percent decrease in the amount of watts per square foot needed to light the plant. Reducing energy consumption by 15 percent of 1990 usage would save $2 million annually and reduce carbon-dioxide emissions from power plants by 69,000 tons. Employee education promotes turning off lights, computers and other machinery when not in use, which alone can save 5 to 10 percent and more than $1 million. Some office areas are equipped with occupancy sensors.

Employees were supplied with an office equipment energy statement to help make informed decisions about when and why to turn off equipment and a hotline to answer questions.

Employee Training

Hallmark created a corporate energy efficiency team that initiated a week-long energy-management workshop for facility maintenance personnel. Topics included electricity, lighting, motors, air compressors and heating and cooling. Employees learned how to gauge the efficiency of heating and cooling equipment and other ways of curtailing energy costs.

Employees participated in a seminar called "The Economics of Environmentalism" at nature sanctuaries and centers near Kansas City where naturalists discussed how to coexist with the environment and annual environmental, health and safety workshops are conducted for more than 100 employees from various facilities and subsidiaries.

Awards

Hallmark offered its first Environmental, Health and Safety Continuous Improvement Awards celebrating the progress toward water-based printing to employees who had worked to phase out solvent-based inks and install water-based inks.

Hoerscht Marion Roussel

Kansas City, Missouri

HMR, an international pharmaceutical company, has eliminated the cotton from bottles of its prescription products in the Untied States, which will save 110,500 pounds of cotton waste or 1,418 miles of cotton wads per year. Originally used to protect tablets from breaking during shipment, the company proved in shipping tests in 1993 that drugs could withstand breakage during shipment without cotton.

Package and Waste Reduction

Child-resistant caps formerly were put on bottles containing 100 tablets, but pharmacists threw the caps away because they were the only ones handling the bottles of 100 doses as they created 14- and 30-day doses of product. HMR converted to standard caps for large-count bottles and saved 194,000 pounds of waste and 8,600 cubic feet of landfill space.

Some tablets have been reduced in size, along with their blister card pack, reducing aluminum use by 11,700 kg per year plus PVC and paper box material.

A Canadian plant programmed its photocopiers to choose two-sided copies as the default rather than one-sided. An education campaign for employees on using the copier has saved 6.7 million sheets of paper in one year for a site of 300 people.

Shrink-wrapped six-packs of bottled tablets make them easy to handle at the point of origin, but the wholesaler breaks the packs to supply one or more bottles to pharmacists. The Puerto Rico plant decided to stop using shrink wrap and saved 1,200 pounds of plastic per month going to the landfill.

Emissions

HMR has reduced some emissions by 66 percent from 1988 to 1992. Reductions are estimated at 95 percent for 1995. The Kansas City facility reduced its air emissions of some compounds from 189,972 kilograms in 1988 essentially to zero in 1992. Better environmental controls and product reformulations have created the decrease.

S.C. Johnson Wax

Racine, Wisconsin

Johnson Wax has a long-standing reputation for environmental sensitivity. The company introduced sustainable carnauba palm leaf harvesting from Brazil to provide wax for the company's products in 1935. It banned CFC propellant use in 1975, three years before they were banned by the U.S. government. The company has commissioned Roper Reports on environmental awareness, behavior and knowledge, and chairman Samuel C. Johnson is a founding member of the Business Council for Sustainable Development and the President's Council on Sustainable Development in the United States. These are the worldwide environmental goals the company established in 1990:

- Phase out the use of specific chemicals and cease formulating new or restaged products with these identified chemical ingredients worldwide.
- Reduce volatile organic compound ratio to total raw materials by 25 percent by the end of 1995.
- Reduce the amount of virgin packaging as a ratio to formula weight by 20 percent by the end of 1995.
- Reduce air emissions, water effluents and solid waste disposal in manufacturing operations by 50 percent as a ratio to total production by the end of 1995.
- Recycle 90 percent of wastepaper, cardboard, plastic, glass and steel materials in manufacturing and office facilities (this objective pertains primarily to North America and Europe, where recycling infrastructure are in place).
- Support educational programs that promote meaningful environmental behavior by the individual.

Eliminating Chemicals

The company has reformulated or eliminated products containing lead, arsenic, CFCs and ethyl compounds.

Emissions

In 1993, 50 products were reformulated to reduce volatile organic compounds. Half of these products are now in production; the rest are nearing completion of reformulation. The company has reduced its overall, worldwide use of VOCs by 17 percent.

Packaging

Johnson Wax uses the virgin packaging-to-formula ratio to measure both source reduction of packaging and increased recycled content. "The virgin packaging amount is calculated by subtracting the recycled amount from the total packaging amount for each material." This measurement also "allows a concentration factor on the formula weight to reflect formula concentrations that deliver more product performance with the same or less packaging." By the end of 1993 it has reduced virgin packaging as a ratio to formula by 14.3 percent worldwide since 1990.

Some Product Innovations

A new electron flytrap called Vector traps flies without high voltage, which eliminates airborne insect and metal particles and bacteria.

Spray can plastic overcaps in products such as Pledge and Raid were reduced in weight by 40 percent, reducing their total annual weight by 2,400,000 pounds.

Packaging redesign for Glade Plug-Ins reduced paperboard by 92,400 pounds annually since 1991.

Reducing Pollution and Waste

In the Netherlands, a costly on-site water treatment operation was eliminated by retrieving, filtering, recycling and reusing water from manufacturing processes in a complete system.

The Waxdale facility in the United States reduced solid waste by 60 percent over 1985 levels. Projections for a total reduction of 90 percent by 1995 will include the 10-year period. To date more than 17,800,000 pounds of waste materials have been diverted through a partnership with Goodwill Industries of southeastern Wisconsin.

The Waxdale facility uses up to 33 percent of its steam power from methane gas from a local landfill to power the plant.

The Brazil facility converts wastewater sludge into brick through an outside operation. These bricks are used in new buildings at the site.

Community Outreach

This company convened an Aerosol Recycling Conference in 1993. It offers annual "We Care for America" and "We Care for Canada" coupon programs in partnership with World Wildlife Fund. It sponsored the Living Planet education curriculum, the Teen America's environmental GPA for environmental literacy, the Global Youth Forum with youth from more than 60 countries, youth programs in Ireland and Australia, and a Spanish translation of Living Planet for distribution to schools in Mexico City.

Laidlaw

Toronto, Canada

Though Laidlaw is a Canadian company, its strong presence in the United States earns a place here. Laidlaw ranks second in the management of hazardous waste and is the third largest manager of solid and biomedical waste and recyclable resources. Laidlaw Waste Systems pioneered the "Blue Box" system in Canada for curbside collection for recyclables, which has since spread to the United States.

The company appreciates the public's increased interest in solid hazardous waste and the growing need for customers to know what happens to their waste after it leaves their premises. For that reason it has become more customer and community attuned and has developed partnerships to solve the solid waste problem.

- The Waste Auditor program provides consumers with an analysis of their waste stream and recommendations to reduce, reuse or recycle the wastes cost effectively. Laidlaw's audit for a hospital, for instance, revealed more than 2,800 kilograms of used cloth gloves disposed of annually. Laidlaw recommended a collection and laundering system, which recovered 80 percent of the gloves and saved $4,000. The company has conducted more than 250 waste audits, including the Toronto Blue Jays' stadium waste after a 1992 World Series game.

- Laidlaw, Pitney Bowes and Purolator Courier collect, transport and recondition toner cartridges. The program could recover nearly 200 tractor-trailer loads of used toner cartridges annually. Pitney Bowes and Laidlaw have also begun work on a project to disassemble used postage meters, copiers and other business machines. Machines are broken down and all usable parts recycled. The program has a 95 percent diversion rate from the waste stream.

- Laidlaw Resources works to market all recyclables collected from its curbside pickups. It finds end markets for glass, newsprint, aluminum, plastics, corrugated cardboard and office paper. To date it has created markets for more than a million tons of recyclables annually, including nearly 400,000 tons of newspapers and magazines.

- Its Earth Academy program educates school children across North America to the importance of the 3 R's: reduce, reuse, recycle.
- Laidlaw's plan for meeting federal regulations for U.S. landfills has been used by the EPA in drafting technical guidance for its own regulations and distributed by the EPA as a model approach to ensuring compliance.
- Laidlaw Waste Systems has created an alternative to the daily sand cover required at landfills. A thin layer of a fiber and stucco-like binder is sprayed on instead, not only covering the waste, but conserving landfill space.
- Laidlaw Environmental Services, with the cooperation of South Carolina's Waterfowl Association, is sponsoring construction of a waterfowl aviary and overnight lodging facility for educational tours at its mined-out property in that state.
- A 10-year restoration project at the west end of Lake Ontario will restore wetland and woodland habitats, provide educational and recreational opportunities and increase the water quality of the Great Lakes. Project Paradise, as it is called, is under the management of the Royal Botanical Gardens.
- Laidlaw's transportation system is investing in alternative-powered vehicles.

McDonald's

Oak Brook, Illinois

McDonald's gained attention in 1990 and 1991 when it switched from styrofoam shells for its hamburgers to paper wrap. It has been reducing waste and making other changes ever since.

Recycling

The company increased its recycled paper content in packaging from 17 percent in 1990 to 45 percent in 1995. Half of the recycled content is postconsumer. The restaurants in the United States use 220,000 tons of recycled packaging each year. It purchases more than $250 million in recycled products of all kinds each year. Many packaging modifications, such as reducing the back flap of french fry cartons, have saved tons of paper.

The company helped to establish the Paper Task Force with Duke University, the Environmental Defense Fund, Johnson & Johnson, NationsBank, Prudential and Time to increase the use of environmentally preferable paper products by the middle of 1995. McDonald's joined the EPA's Waste Wi$e Program.

McRecycle USA offers vendors of recycled materials a potential market of $100 million of recycled products annually for new and remodeled restaurants. McDonald's will consider any materials that meet specifications, standards of quality and competitive prices from carpeting to utility carts for its restaurants. McDonald's has purchased more than $200 million worth of recycled products annually since 1990. Vendors can register their products and supplies of recycled products with the McRecycle Registry Service. On average each restaurant spends $22,000 annually on recycled products. The company annually has spent more than $600 million on recycled paper for use in napkins, drink trays, trayliners and Happy Meal boxes annually since 1990.

McDonald's reduced the volume of sandwich packaging by 90 percent, which enabled it to reduce the size of its shipping boxes and reduced energy consumption as well. The company has also reduced the size of its napkins, straws, Happy Meal bags and hot cup lids. McDonald's uses unbleached paper products or more benign bleaching processes wherever possible.

Behind the counter, McDonald's uses Coca-Cola drink syrup in reusable containers instead of cardboard, which cuts millions of pounds of packaging yearly. A corrugated cardboard recycling program turns cardboard into new boxes and carry-out bags. Hamburger buns now come in reusable plastic bakery containers.

Energy

McDonald's has joined the Green Lights program sponsored by the EPA and is replacing 40-watt lamps with 32-watt lamps that generate more light. The program is handled by Sylvania, which outfits the stores during nonrush business hours for a cost from $1,800 to $5,500 per store. Savings per store averages about $1,300, which means the program pays for itself in one to three years. The company plans to convert 90 percent of McOpCo restaurants by 1998.

Awards

In 1990 McDonald's joined with the Environmental Defense Fund to reduce waste materials and create a model approach for other companies. The task force earned the President's Environmental and Conservation Challenge Award in 1991. By 1993 McDonald's had completed more than 80 projects and initiatives—twice the number in the original plan.

Community Outreach

McDonald's has joined with Clemson University, Conservation International, the Tropical Science Center in Costa Rica and the Sustainable Development Foundation in Panama to provide $4 million toward conserving some of Costa Rica's rainforest. The program is called AMISCONDE (Amistad Conservation and Development Initiative) and seeks to halt deforestation and restore the land while providing income for local farmers. In conjunction with Conservation International, McDonald's developed a 25-minute video called *The Rain Forest Imperative* to help students understand rainforest preservation issues. The video has been made available to classrooms around the world. McDonald's has also developed a classroom poster with the World Wildlife Fund to provide information on tropical rainforests.

The company publishes *WEcology Magazine,* a 16-page magazine developed with World Wildlife Fund for grades 6–10, featuring articles on the environment.

3M Corporation

St. Paul, Minnesota

In 1975 the company created the Pollution Prevention Pays (3P) program to prevent pollution at the source in products and manufacturing processes, a strategy that would avoid having to remove pollution after it is created. This effort makes 3M one of the most environmentally mature companies; it is one of the very few companies that talks about committing to "sustainable development," a goal that few can envision, much less reach.

Quantifiable Results

3M's worldwide savings 1975–1993, as reported March 1994, are estimated at $710 million and fall into these categories:

	United States	International
Air pollutants	158,000 tons	19,000 tons
Water pollutants	16,800 tons	1,400 tons
Wastewater	2 billion gallons	700 million gal
Sludge/solid waste	433,000 tons	26,000 tons

$577 million has been saved in the United States and $133 million in international operations from more than 4,000 projects.

Energy

An energy management program to reduce energy use throughout facilities in the United States has cut energy use by 50 percent per unit of production and associated air emission of certain pollutants by more than 25 million pounds per year and carbon dioxide by more than 3.9 billion pounds per year.

A Commute-A-Van program for employees has saved 3 million gallons of gasoline and eliminated more than 60 million pounds of air emissions since 1973.

In resource recovery programs, the company has saved more than $120 million from its resource recovery operations since 1985. In 1990 alone, more than 82 million pounds of materials were recovered, reused or sold. Its main strategies to prevent pollution include:

- Product reformulation
- Process modification
- Equipment redesign
- Recycling and reuse of waste materials

The company's goals are to cut "all hazardous and nonhazardous releases to the air, water and land by 90 percent and to reduce the generation of waste by 50 percent by the year 2000 from a base line of 1990." Ultimately it wants to reduce emissions to negligible levels.

Emissions

An updated 3P program, 3P Plus, focuses on scientific research and increased attention to pollution prevention. 3M scientists have reduced the use of solvents in the manufacture of many products; this research continues. In 1994 the company installed advanced pollution control technology to reduce air emissions in facilities worldwide. Total cost was $175 million. This Air Emissions Reduction Program has reduced emissions 70 percent.

The Hilden, Germany, 3M plant had to dispose of 110 tons of cleaning solvent and use 800 staff hours to clean its solvent storage tanks yearly. By developing an automated, fully enclosed sedimentation cleaning and recovery system, the plant now recovers solvent from the cleaning solution while eliminating worker contact with the chemicals. Less solvent now needs to be transported for disposal.

A water-based coating was substituted for a solvent solution coating at 3M's Northridge, California, plant, eliminating the need to install additional pollution control equipment and preventing more the 48,000 pounds of air pollution a year. The change cost $60,000.

One story about early planning and communication when expanding or developing new sites is a lesson in good management: An office and laboratory complex in Austin, Texas, was planned for expansion but would run into habitat for the endangered golden-cheeked warbler. 3M representatives in Austin talked with city officials and local environmental groups. After two years of negotiations, 3M agreed to provide 215 acres of land close to the golden-cheeked warbler habitat in exchange for taking 12 acres of existing warbler habitat. The company worked with the U.S. Fish and Wildlife Service and others to develop a long-range conservation plan to help protect the warbler. The company believes early communication and willingness to create a win-win solution avoided unpleasant confrontations later.

Awards

The company has won numerous awards, some of which include the National Energy Resources Organization Award, 1980; the United Kingdom Commendation, 1983; the 1st Annual World Environment Center Award, 1984; governors' awards from Minnesota and Alabama in 1984 and 1985, respectively; America's Corporate Conscience Award from the Council on Economic Priorities, 1988; the National Wildlife Federation Corporate Conservation Council's Environmental Achievement Award, 1989.

Monsanto

St. Louis, Missouri

As a result of gathering data for its first toxic release inventory in 1987, the company set a goal to reduce toxic air emissions worldwide by 90 percent by the end of 1992. Through 1993 the company had reduced its targeted air emissions by 90 percent worldwide, 81 percent for the United States. Its next goal is not to slip back and to reach zero. Most of the reductions came from consolidating or shutting down inefficient processes. Monsanto employs about 30,000 people worldwide and makes such products as NutraSweet, nylon fibers for Wear-Dated carpet, Roundup herbicide and pharmaceuticals.

Emissions

The company committed more than $100 million to reduce air emissions over five years. Monsanto employees implemented more than 250 projects in achieving the air emissions goal. Most of the gains came from installing pollution controls, consolidating or shutting down inefficient units and modifying processes or recovery practices to prevent toxic chemical emissions. Worldwide emissions were reduced from 60 million pounds per year to six million pounds. U.S. emissions were down from 17 million pounds per year in 1987 to three million pounds.

Installing two large-scale biofilters in Massachusetts and Michigan plants helped achieve the 90 percent goal. The biofilters resemble compost beds the size of basketball courts. Made of wood chips and naturally occurring soil microorganisms, the biofilters destroy chemical emissions in dilute air streams at these two plants.

Engineers at the LaSalle plant near Montreal, Canada, tripled their manufacturing capacity while virtually eliminating liquid waste. The plant makes Scripset resins, which give a high-quality surface to paper for writing, printing and copying. Flow-through systems were replaced by highly efficient closed-loop systems. Now 90 percent of the xylene and ammonia formerly discharged to wastewater is captured and reused. The remaining 10 percent is treated.

A new process at its plant in Luling, Louisiana, eliminates approximately 10 million pounds of waste annually. The process eliminated solid and liquid wastes in the production of disodium iminodiacetate, used in the manufacture of Roundup herbicide.

Community Outreach

The Pensacola, Florida, plant has created a 1,500-acre wildlife sanctuary on plant property with a 500-acre wildlife area and a place where the public can view waterfowl.

Employees at the Puerto Rico plant installed chlorinated drinking water units and repaired existing ones to protect the water quality of more than 200 communities with a combined population of 100,000. They also trained residents how to operate equipment and handle chlorine containers. This activity is part of a mutual association with local businesses and the U.S. EPA to improve water quality in Puerto Rico.

JC Penney Company

Dallas, Texas

JC Penney adopted its "Principles on the Environment" in 1991 and has undertaken a program to eliminate waste and increase energy efficiency. It has an Environmental Affairs Committee, which leads in implementing the principles.

Package Reduction

Nearly 500 million hang tags and price tags were changed in the beginning of 1993 to incorporate recycled paper, eliminate unnecessary tags or reduce the size and convert to water-based varnishes.

Catalog tie boxes have been modified to a corrugated box with minimal printing and no film window to increase recyclability. Gift, jewelry and most shoe boxes consist of more than 90 percent recycled paperboard, which contains 80 percent postconsumer waste fibers. Packaging engineers now include specific weight and volume data, which will be used to track progress. Packaging suppliers are now asked to submit alternative packaging solutions in addition to quoting on the approved or existing specifications. All buyers have guidelines for evaluating and substantiating environmental claims for merchandise, packaging and labeling to assure that the company complies with all legal requirements and communicates a clear, consistent message to the public.

Recycling

JC Penney's goal is to reduce its solid waste entering the municipal landfills by 60 percent. Recycling programs are in place at all six catalog distribution centers, 14 telemarketing centers and 15 catalog-outlet stores. Together these centers recycle more than 15,000 tons annually. The financial services office in Plano, Texas, recycles about 30,000 pounds of paper each month.

About 70 percent of the company's solid waste is corrugated cardboard, which it has bailed and recycled for more than five years. Its second-largest source of solid waste, plastic, is also being reduced. JC Penney is working with vendors to reduce and eliminate plastic packaging, disposable hangers and other sources of plastic. Some stores are testing if returned, used catalogs can be recycled. If the tests are successful, the program will go companywide.

Energy

The company has reduced its electrical energy consumption by 50 percent over the last decade through heating and air conditioning system improvements and high-efficiency lighting. A capital investment of $42 million has reduced annual electrical consumption by 400 million kwh, for an estimated savings of 540 million pounds of coal or 43 million gallons of oil.

JC Penney has adopted an efficient form of fluorescent lighting that will reduce lighting energy consumption by more than 50 percent in 30.2 million square feet of store space. The company is working with the EPA in their Green Lights Program.

Procter & Gamble

Cincinnati, Ohio

P&G's Total Quality Environmental Management (TQEM), principles followed by the company's manufacturing operations, "views waste and all forms of pollution as quality defects to be corrected." TQEM promotes the continuous improvement cycle that helps managers set goals, initiate changes and track progress in environmental compliance, safety, pollution prevention and waste reduction.

Waste Reduction

The company has achieved a 49 percent worldwide reduction per production unit in waste since 1990 and it reuses or recycles 60 percent of the wastes generated. A Design-Waste-Out program identifies and eliminates waste from the design through the manufacturing stages of a product's life.

Some examples of waste reduction in packaging include toothpaste marketed in Europe in stand-up, all-plastic laminate tubes without a carton. This redesign has reduced toothpaste packaging by 207 tons per year. Cartonless Crest toothpaste in Canada represents a 30 percent reduction in packaging, equivalent to 170 tons a year. Outer cartons for Secret and Sure underarm products have been eliminated in the United States, a reduction that saves 80 million cartons a year or 3.4 million pounds of paperboard. Crisco, the U.S. cooking oil, has a new bottle, which uses 30 percent less material than the previous one. The new bottle will save some 2.5 million pounds of plastic and 1.3 million pounds in corrugated shipping containers per year.

P&G has created refill packages for more than 57 brands in 22 countries around the world. Conventional packaging typically costs more and adds up to 70 percent more solid waste.

The company's plant in Worms, Germany, which manufactures products, such as Ariel, Dash and Sanso detergents, created a new waste-management operation. The system received presorted solid waste, then sorted it again and sold the materials. This program combined with other waste reductions lowered the total waste stream by 85 percent. This particular improvement saved $4.3 million in disposal fees in fiscal

1992–1993 and created the same savings in the value of recovered raw materials and packing materials.

Life-Cycle Assessment

Life cycle inventories have been conducted on diapers, all-purpose cleaners, laundry detergents and other products. This analysis helps shape a comprehensive environmental management strategy. In 1987 the company constructed a $1.5 million experimental stream facility on the little Miami River outside of Cincinnati. The stream facility allows researchers to investigate the effect of a product ingredient on natural ecosystems.

Energy Savings

One example of energy savings is a coal-burning manufacturing plant in the Czech Republic acquired in 1991 and refitted to bring the plant up to P&G's safety and environmental standards. Coal-fired boilers were replaced with gas burners, eliminating 13 million pounds of boiler ash and 221,000 pounds of fly ash per year. Sulfur dioxide emissions were reduced by 1.3 million pounds annually. The total energy conservation program reduced overall fuel consumption by 43 billion BTUs per year.

Wastewater

The Lima, Ohio, plant wanted to reduce the amount of wastewater effluent by 50 percent over 12 months with a continuous improvement goal of zero discharge. One of the changes was the installation of an ultrafiltration unit and equipment redesign. Discharges have dropped by 70 percent to date.

Recycling

P&G has used recycled fiber in paperboard packaging for more than 20 years. More than 80 percent of its paper packaging is now made from recycled materials. P&G was the first manufacturer in the United States to market a product, Spic and Span Pine liquid cleaner, in a bottle made from 100 percent post-consumer recycled plastic. Its detergent bottles and diaper packages contain at least 25 percent postconsumer plastic.

Composting

Acting on the knowledge that 30 to 50 percent of municipal solid waste can be recovered by composting, P&G has committed $20 million worldwide to help establish composting in communities in Korea, Belgium, France and The Netherlands. In the United States P&G joined with the National Audubon Society in 1992 to test the feasibility of having consumers separate compostable and recyclable materials for curbside pickup. Results indicate that composting and recycling together could divert as much as 70 percent of household refuse.

Community Outreach

P&G helped to establish GEMI (Global Environmental Management Initiative), an organization that provides a forum for environmental management discussion. Through GEMI, P&G works with other companies in environmental auditing and pollution prevention.

The company is a member of the Corporate Conservation Council of the National Wildlife Federation, which brings corporate executives and environmentalists together. In 1991 the company helped fund the purchase of 20,000 acres of prairie wetlands in South Dakota.

Its educational programs include *Planet Patrol* for grades 4–6, *Medicine Man* for grades 11–12, and *Decisions About Product Safety,* information about chemical safety. *Our Weakening Web,* a traveling exhibit, stresses the importance of ecological relationships and citizen action to provide solutions. The display will tour 33 cities in the United States and Canada through 1998.

Sun Company, Inc.

Philadelphia, Pennsylvania

Sun is the largest independent U.S. refiner and marketer of oil. Sunoco markets gasoline and other petroleum products through 4,500 service stations located mainly in the northeastern United States. A-plus Mini Market has 540 convenience stores. The company also produces crude oil and natural gas internationally and is the major owner of Suncor Inc. in Canada.

Sun published its first environmental report in 1992 and endorsed the CERES Principles in 1993, which has resulted in new partnerships with environmental, religious groups and social investors. The company agreed on the need for cost-effective solutions that would support Sun's continued success and discussed making long-term improvements in its environmental performance.

Emissions

Sun is involved with the EPA's 33/50 program and expects to achieve its 50 percent reduction target based on 1988 levels by December 1995. In 1988 the company emitted 6.8 million pounds per year. In 1992 emissions had dropped to 4.9 million pounds. Projected expenditures to achieve the 1995 goal are $2.5 million.

Alternative Fuels

Sun has developed and is test marketing low-emission fuels:

- Liquefied petroleum gas marketed as Sunoco "AutoPropane"
- M85, a blend of 85 percent methanol and 15 percent gasoline
- E85, a blend of 85 percent ethanol and 15 percent gasoline
- Compressed natural gas

The company is reformulating gasoline for the Clean Air Act's 1995 implementation date. Sun has lowered the amount of benzene in its reformulations to reduce toxic emissions. Added oxygenates reduce hydrocarbon and carbon monoxide emissions.

The company opened a new facility in 1994 in Mon Belview, Texas, which produces methyl tertiary butyl ether (MtBE), a primary oxygenate for gasoline. The plant will be equipped with the latest energy-conservation technology and will allow Sun to meet its oxygenate requirements in 1995 for reformulated gasoline. The plant can produce 12,600 barrels a day of MtBE.

Recycling

More than 7,000 tons of solid catalysts, used to help convert crude oil into fuels, had been dumped in landfills annually. Sun has developed a way to recycle this material through cement plants and steel mills instead, saving both raw materials and landfill space.

The company developed a system to capture benzine vapor emitted from barge shipments to chemical plant customers and route it to the refinery to burn as fuel. Another vapor-recovery unit is planned to recycle vapors from gasoline loaded on barges and tankers.

Target

Minneapolis, Minnesota

Target created an environmental department at its Minneapolis headquarters in 1993. The company formed eco-teams with one eco leader per floor. These teams motivate everyone to recycle and act as a sounding board between the employees and the environmental department.

Energy

New stores are constructed with lights that use one-third less energy, and more efficient lighting is installed in existing stores. Outdated but working fluorescent lights are donated to churches, schools and prisons, where they can be reused for as long as two more years.

Recycling

Paper and plastic shopping bags are made with recycled content, without sacrificing strength and quality. Sunday circulars are printed on recyclable paper with some recycled paper content.

Target now uses a 100 percent paperless system on domestic purchase orders. This involves 2.7 million transactions and keeps 40 tons of paper out of the landfills. Target is the first discount retail in the United States to use this system.

The company repairs and reuses wooden shipping pallets rather than throwing them away. All store signs are printed on recycled stock and the stores use recycled computer paper. Target recycled an estimated 38 tons of computer paper in 1993.

All the stores recycle 75 percent of their waste and are working on the remaining 25 percent. Their national cardboard-recycling program keeps more than 100,000 tons of cardboard out of landfills.

The plastic wrap recycled each year weighs as much as 1 million pounds. The company reuses it in garbage bags.

Community Outreach

Target has returned 5 percent of pretax dollars to the stores' communities for environmental, arts and social action activities and organizations since 1962.

Target and Kodak raised $300,000 to help preserve national parks and produce the *This Land Is Your Land* campaign. Robert Redford and Vice President Al Gore gave a five-minute radio address, asking everyone to get involved in preserving our parks.

Target has donated merchandise to nonprofit organizations since its early years. Opened pet food and cat litter go to the Humane Society; samples of clothing, furniture, toys, etc., go to Goodwill Industries; opened cans of paint to Habitat for Humanity.

Walmart Stores, Inc.

Bentonville, Arkansas

Wal-Mart announced its commitment to environmental quality in 1989. Manufacturers and vendors were challenged to improve products and packaging, making them more environmentally responsive. Wal-Mart auto centers take used motor oil and batteries in their recycling programs. All stores recycle their cardboard and plastic shopping bags. Each store has a Green Coordinator who answers environmental questions and educates customers to green products. The company participates in Earth Day programs, local parades, tree plantings, information booths and in-store videos and fund-raising activities for local environmental groups.

Demonstration Store

Wal-Mart opened its environmental demonstration store in Lawrence, Kansas, in 1993. The store contains an environmental education center with displays, videos, slides and narration; a community classroom; a recycling center and skylights with photo sensors mounted at the base of the skylight wells, which measure the amount of daylight entering the building and automatically adjust the level of the light. They work with the fluorescent lights, which are equipped with solid-state electron ballasts. Glazing of the skylights allows more daylight into the building compared to other skylights. Solar-optic films on some skylights distribute the daylight evenly in a radius from the center of the unit. The roof itself and the ceiling structure is an engineered wood beam system. Wood was chosen because it requires substantially less fossil fuel energy to harvest and manufacture and is a renewable resource.

Temperature Control

The heating and cooling system uses no refrigerants that deplete the ozone layer. The system was designed to use 100 percent fresh air for building ventilation. An ice storage system works this way: The building's air conditioning system cools the building during operating hours and makes ice during the hours when the store is closed. The ice is melted during the day to reduce the peak air conditioning demand.

Water Usage

All storm water from the parking lot and roof as well as sinks and drain water are collected in a gray-water system, treated, stored in a detention pond and reused as irrigation water for landscaping. The landscaping itself requires low water and maintenance. Much of it is native to the local climate. Low-water fixtures are used to reduce the consumption of city water.

Recycled Materials

Bumper blocks, shopping cart corrals, directional signs are all made from recycled materials.

Package Recovery

Customers can leave packages from purchased products at the front of the store before leaving.

Recycling Center

The recycling center offers recycling bins for plastics. It also accepts tin, aluminum, glass, newspapers, corrugated cardboard and other products. Revenues are donated to local charities.
Plans are in the works for other similar stores.

Xerox Corporation

Rochester, New York

Xerox has been environmentally conscious since the 1960s when the company reclaimed metals from photoreceptor drums and saved money by using recycled metals. Reclaiming metals from used photoreceptors now annually yields 800,000 pounds of nickel, 600,000 pounds of aluminum and 160,000 pounds of selenium. Recycling scrapped parts represents an annual recovery value of $1.8 million.

Xerox also practiced product life-cycle management early as it converted old machines into functional components for reuse. In the 1970s the company started energy conservation programs at all its major sites and an employee car pool program in major metropolitan areas. Xerox introduced both energy-saving models and two-sided copying in copiers and printers during that time.

In the 1980s Xerox improved its service vehicle fuel consumption from 13 to 20 miles per gallon and performed an environmental assessment of all major Xerox sites and operations. As a result, all underground storage tanks were replaced with above-ground tanks with double lining for maximum containment of materials. The use of solvents in remanufacturing processes was reduced by 90 percent by changing to citrus-based cleaners and soap and water. Company policy states that the environment and the health and safety of its employees should be protected over economic considerations.

Design for the Environment

Since 1990 the company has increased packaging and newspaper recycling at all sites, started using methanol-powered service vehicles, reduced the use of ozone-depleting chemicals in all processes and replaced chemical solvents with water-based and organic cleansers in manufacturing. Xerox calculates each product's total cost, including energy, from harvesting raw materials to final disposal. In 1993 the company phased out CFCs and trichlorethane from the manufacturing process and converted toner bottles to 100 percent recycled plastic. Xerox's goal is to create products that produce virtually no waste during manufacturing, use or at the end of their initial life cycles.

Environmental Leadership Program

In 1990, Xerox established the Environmental Leadership Program to champion programs in resource conservation and waste reduction. Waste reduction programs have increased recycling rates at Xerox's largest facilities from 34 percent on average in 1990 to an average approaching 73 percent today. Source reduction and resource conservation figure largely in this, affecting both operations and design and manufacturing methods.

Product Life Cycles and Design for Disassembly

Environmental design includes helping designers choose nontoxic materials such as thermoplastics and metals for manufacturing parts. These materials wear well and adapt to reuse and recycling. The design includes designing for disassembly as well. This practice improves equipment remanufacturability and facilitates routine service and maintenance. Simplified components need fewer repairs and are easier to replace, lowering service costs and downtime. Remanufactured components meet the same standards as new ones and come with the same satisfaction guarantee.

Customers can ship used copy cartridges back to Xerox in the original packaging. Xerox pays for the shipping and remanufactures the cartridges. Its dry ink and toner containers are now made from as much as 100 percent postconsumer recycled plastic. The company also offers copier/printer papers with postconsumer content up to 50 percent.

Awards

Xerox has received numerous awards, including the World Environment Center's 1993 Gold Medal for International Corporate Environmental Achievement, the National Environmental Development Association 1992 Honor Roll (U.S.) and Brazil's Selo Verde Award. Products have qualified for environmental quality labels such as EPA's Energy Star, Germany's Blue Angel and Canada's Environmental Choice. For product quality the company has earned "the Grand Slam" of quality awards: The Malcolm Baldridge National Quality Award in the United States, the Deming Prize in Japan and the First Annual European Quality Award.

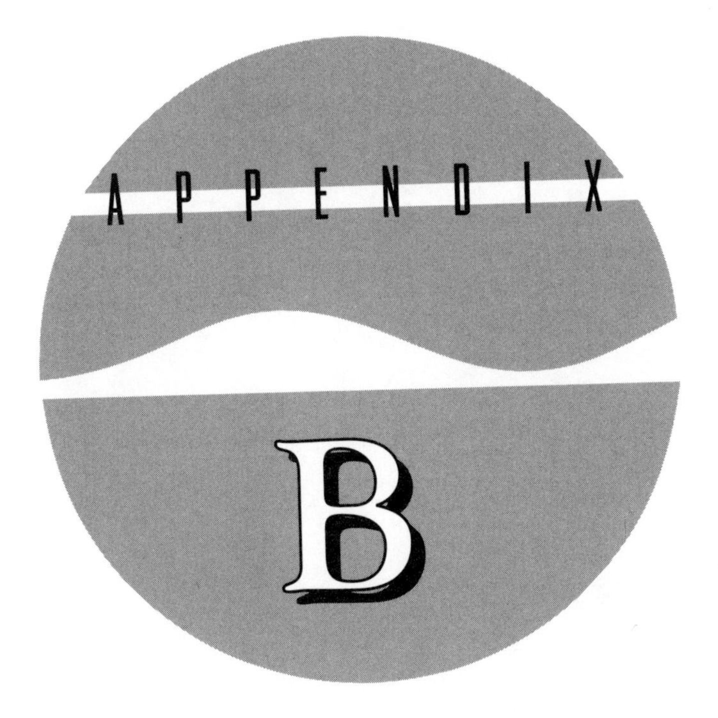

The Coalition for Environmentally Responsible Economies

The CERES Principles

Introduction

By adopting these principles, we publicly affirm our belief that corporations have a responsibility for the environment and must conduct all aspects of their business in a manner that protects the earth. We believe that corporations must not compromise the ability of future generations to sustain themselves.

We will update our practices constantly in light of advances in technology and new understandings in health and environmental science. In collaboration with CERES, we will promote a dynamic process to ensure that the principles are interpreted in a way that accommodates changing technologies and environmental realities. We intend to make consistent, measurable progress in implementing these principles and to apply them to all aspects of our operations throughout the world.

Protection of the Biosphere

We will reduce or eliminate the release of any substance that may cause environmental damage to the air, water, earth or its inhabitants. We will safeguard all habitats affected by our operations and protect open spaces and wilderness, while preserving biodiversity.

Sustainable Use of Natural Resources

We will make sustainable use of renewable natural resources, such as water, soils and forests. We will conserve nonrenewable natural resources through efficient use and careful planning.

Reduction and Disposal of Wastes

We will reduce and, where possible, eliminate waste through source reduction and recycling. All waste will be handled and disposed of through safe and responsible methods.

Energy Conservation

We will conserve energy and improve the energy efficiency of our internal operations and of the goods and services we sell. We will make every effort to use environmentally safe and sustainable energy sources.

Risk Reduction

We will strive to minimize the environmental, health and safety risks to our employees and the communities in which we operate through safe technologies, facilities and operating procedures, and by being prepared for emergencies.

Safe Products and Services

We will reduce and, where possible, eliminate the use, manufacture or sale of products and services that cause environmental damage or health or safety hazards. We will inform our customers of the environmental impacts of our products or services and try to correct unsafe use.

Environmental Restoration

We will promptly and responsibly correct conditions we have caused that endanger health, safety or the environment. To the extent feasible, we will redress injuries we have caused to persons or damage we have caused to the environment and will restore the environment.

Informing the Public

We will inform in a timely manner everyone who may be affected by conditions caused by our company that might endanger health, safety or the environment. We will regularly seek advice and counsel through dialogue with persons in communities near our facilities. We will not take any action against employees for reporting dangerous incidents or conditions to management or to appropriate authorities.

Management Commitment

We will implement these principles and sustain a process that ensures that the board of directors and chief executive officer are fully informed about pertinent environmental issues and are fully responsible for environmental policy. In selecting our board of directors, we will consider demonstrated environmental commitment as a factor.

Audits and Reports

We will conduct an annual self-evaluation of our progress in implementing these principles. We will support the timely creation of generally accepted environmental audit procedures. We will complete the CERES Report annually, which will be made available to the public.

Disclaimer

These principles establish an ethic with criteria by which investors and others can assess the environmental performance of companies. Companies that endorse these principles pledge to exceed the requirements of law. The terms "may" and "might" in principles one and eight are not meant to encompass every imaginable consequence, no matter how remote. Rather, these principles obligate endorsers to behave as persons who are not governed by conflicting interests and who possess a strong commitment to environmental excellence and to human health and safety. These principles are not intended to create new legal liabilities, expand existing rights or obligations, waive legal defenses, or otherwise affect the legal position of any endorsing company and are not intended to be used against an endorser in any legal proceeding for any purpose.

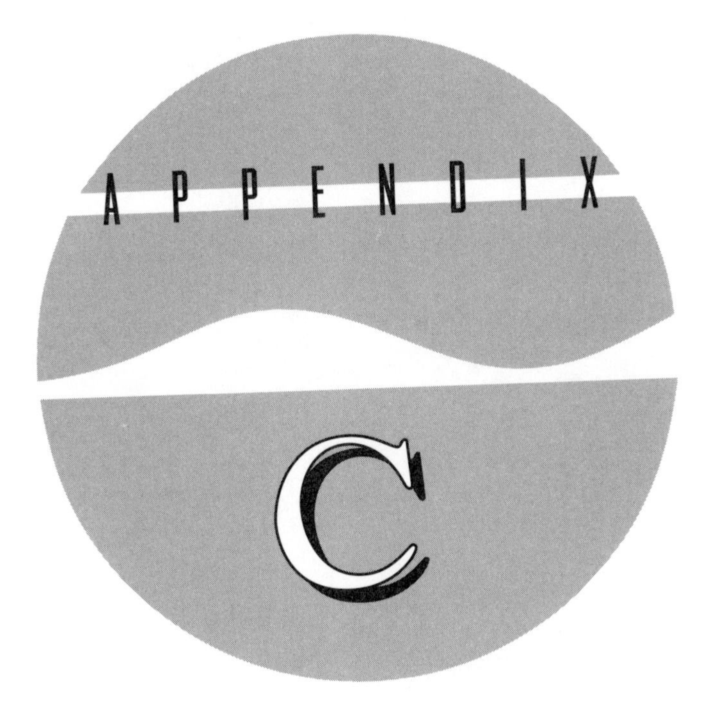

A checklist for creating a sound environmental ethic in your organization

IN SEARCH OF ECO-EXCELLENCE

Manufacturing

- ❍ Could the product be cleaner?
- ❍ Could it be more energy efficient?
- ❍ Could it be more intelligent?
- ❍ How long will it last?
- ❍ What happens to it when its life ends?
- ❍ Is there a secondary environmental market?
- ❍ Can it be designed for reusability and recycling?
- ❍ Can it be easily repaired or its life extended through regular maintenance?
- ❍ Is it manufactured with least toxicity? Have you work toward substituting benign substances?
- ❍ Can alternative materials be used that put less pressure on exotic or fragile ecosystems?

- ❍ Does the process meet and exceed OSHA and EPA standards?
- ❍ Has a life-cycle analysis of the true cost of the product been done?

Marketing

- ❍ Have staff been through ecological sales training?
- ❍ Does the green market campaign come from an authentic position of real change?
- ❍ Green customer service policy—repairability, interchangeability, life-cycle-analysis cost, secondary life, final disposal—is this part of the sales package?
- ❍ Is the company developing positive links with environmental groups?
- ❍ Do communications strategies emphasize environmental aspects?
- ❍ Do you publish an environmental report showing progress in eliminating pollution and waste?
- ❍ Has the company researched the effects of green issues on the firm?

Innovation

- ❍ What are green product characteristics?
- ❍ Do they contribute to lessening global environmental problems? Are they, for instance, CFC or CO_2 free?
- ❍ Are they manufactured from renewable resources?
- ❍ Can they be disposed of safely, reduced and/or reused?
- ❍ Are they manufactured from locally obtainable materials to minimize transport costs?
- ❍ Are they designed to satisfy a genuine human need?

Staff Development

- ❍ How can you increase environmental awareness?
- ❍ What part does the environment play in instruction programs?
- ❍ Could you develop a conservation group as part of the social and professional development activities?
- ❍ Are employees encouraged to participate in the local environmental community?
- ❍ Is environmental responsibility seen as a natural out-growth from diversity, human resource potential and personal responsibility?
- ❍ Are employees recognized for environmental goals, waste reduction and pollution savings?
- ❍ Are cross-department and top-to-bottom teams encouraged in problem solving?

Packaging

- ❍ Is the package necessary?
- ❍ Are any of the components of the packaging unnecessary?
- ❍ Is the package biodegradable, photodegradable, natural, reusable or recycled?
- ❍ Does it exceed its requirements for storage, transport and use?
- ❍ Does it have to be packaged in individual blister packs?
- ❍ Can it be packaged in another way entirely?
- ❍ Is it labeled according to latest guidelines?

Reasonable Packaging Guidelines

- ❍ Does it meet product protection throughout the distribution chain without picking up unnecessary packaging?
- ❍ Are you working with vendors to decrease throwaways?
- ❍ Does shelf appeal provide a final incentive to buy something as a green product?
- ❍ Does environmental safety guard against future environmental hazards?

International Markets

- ❍ Are the workers treated as fairly as in the home country?
- ❍ Are the same environmental standards met as in the home country?
- ❍ Are local customs protected rather than destroyed?
- ❍ Are natural ecosystems restored after raw materials are extracted? Used selectively? Certified as managed for the environment?

Community Outreach

- ❍ Does the company return to the earth what it has taken from it?
- ❍ Does the company honor labor commitments and quality-of-life issues?
- ❍ Does the company do all it can to operate within environmental safeguards?
- ❍ Is the company involved in community support projects?

Corporate Headquarters Philosophy

- ❍ Do corporate objectives include ethical and environmental statements?
- ❍ Has an environmental policy been set?
- ❍ What objectives have been established?
- ❍ Is there board-level responsibility for the program?
- ❍ When considering acquisitions, are environmental impact assessments completed?
- ❍ Has the company completed a social responsibility and environmental audit?
- ❍ Are environmental information systems integrated into management information systems?

BIBLIOGRAPHY

History Timeline

The Bottom Line of Green Is Black: Strategies for Creating Profitable and Environmentally Sound Business, Tedd Saunders, and Loretta McGovern, Harper, San Francisco, 1993.

Dwellers in the Land, Kirkpatrick Sale, Sierra Club Books, San Francisco, 1985.

Ecology, Steve Pollock, Dorling Kindersley, London, 1993.

An Encyclopedia Dictionary of American History, Howard L. Hurwitz, Washington Square Press, New York, 1970.

The Encyclopedia of the Environment, Ruth Eblen and William R. Eblen, Houghton Mifflin, Boston, 1994.

The Encyclopedia of Environmental Studies, William Ashworth, Facts on File, New York, 1991.

Entropy: Into the Greenhouse World, Jeremy Rifkin, Bantam, New York, 1989.

A Forest Journey: The Role of Wood in the Development of Civilization, John Perlin, Harvard University Press, Cambridge, Massachusetts, 1989.

A Green History of the World: The Environment and the Collapse of Great Civilizations, Clive Ponting, St. Martin's Press, New York, 1991.

Indian Givers: How the Indians of the Americas Transformed the World, Jack Weatherford, Fawcett Columbine, New York, 1988.

Living in the Environment: An Introduction to Environmental Science, 6th edition, G. Tyler Miller, Jr., Wadsworth Publishing Company, Belmont, California, 1990.

New England Journal of Medicine, 329:24, December 9, 1993, "An Association between Air Pollution and Mortality in Six U.S. Cities," pp. 1753+.

The Norton History of the Environmental Sciences, Peter J. Bowler, Norton, New York, 1992.

The One Hundred: A Ranking of the Most Influential Persons in History, Michael H. Hart, Carol Publishing, New York, 1993.

The Potato Garden: A Grower's Guide, Maggie Oster, Harmony Books, New York, 1993.

A Sand County Almanac, Aldo Leopold, Ballantine, New York, 1966.

A Timetables of Inventions and Discoveries, Kevin Desmond, M. Evans and Company, Inc., New York, 1986.

The Timetables of Science: A Chronology of the Most Important People and Events in the World of Science, Alexander Hellemans and Bryan Bunch, Simon and Schuster, New York, 1988.

The Timetables of History: A Horizontal Linkage of People and Events, Bernard Grun, Simon and Schuster, New York, 1982.

Ecology of Business

Beyond Compliance: A New Industry View of the Environment, Bruce Smart, World Resources Institute, Washington, D.C., 1992.

The Business Environmental Handbook: How You Can Profit by Being Environment-friendly, Martin D. Westerman, Oasis Press, Grants Pass, Oregon, 1993.

Changing Course: A Global Business Perspective on Development and the Environment, Stephen Schmidheiny with the Business Council for Sustainable Development, MIT Press, Cambridge, Massachusetts, 1992.

The Ecology of Commerce: A Declaration of Sustainability, Paul Hawkin, HarperCollins, New York, 1993.

Earth in the Balance, Ecology and the Human Spirit, Al Gore, Houghton Mifflin, Boston, 1992.

The E Factor: The Bottom-Line Approach to Environmentally Responsible Business, Joel Makower, Tilden Press, 1993.

Environmental Leadership: Developing Effective Skills and Styles, Joyce K. Berry, John C. Gordon, Island Press, Covelo, California, 1993.

From Ideas to Action: Business and Sustainable Development, The ICC Report on the Greening of Enterprise, Jan-Olaf Willums and Ulrich Goluke, ICC Publishing, 1992.

Going Green: How to Communicate Your Company's Environmental Commitment, E. Bruce Harrison, Business One Irwin, Homewood, Illinois, 1993.

Green Gold: Japan, Germany, the United States, and the Race for Environmental Technology, Beacon Press, Boston, 1994.

Green Marketing: Challenges and Opportunities for the New Marketing Age, Jacqueline A. Ottman, NTC Business Books, 1993.

Practical Guide to Environmental Management, Frank B. Friedman, et. al. Environmental Law Institute, Washington, D.C., 1993.

State of the World Reports, Worldwatch Institute Report on Progress Toward a Sustainable Society, Norton, New York. Yearly.